TABLE OF CONTENT

MW00898567

COPYRIGHT & USAGE:

WHAT IS A STEM LAB?

STEM is an acronym for Science, Technology, Engineering, and Mathematics. Recent shifts in education have favored these subjects, primarily because we have a shortage of workforce in these particular areas, which is really quite sad for a number of reasons. These can be some of the most interesting things to study in school, provided they are taught in a fun, interesting, and hands-on fashion. They also lead to some of the best-paying technical jobs, too!

All of the labs within promote these 4 fields. You will find a strong emphasis on designing a project, testing it, measuring the results, and reflecting upon what worked and did not work. The projects are also labeled at the bottom with a series of categorical tags, and they've been sorted into 12 different categories, so you can find similar projects to work on, allowing students to build upon prior knowledge gained in doing these projects! Of course, you can do them in any order you wish, but it can be fun to do a set of similar projects.

Many of the details for each project have deliberately been left vague. It is important for students to design and work toward a goal, rather than have the process spelled out for them. This is a major element of the engineering-design process. However, since these are designed for an educational setting, optional grading suggestions have been added. For fun and homeschool use, discard these recommendations or make up your own grading system to suit your purpose.

ABOUT THIS BOOK

This book contains all of the labs from the 3 original volumes of the series: **50 STEM Labs**, **50 More STEM Labs**, and **50 New STEM Labs**. Some have been edited for clarity and minor error fixes. They have also been resorted into 12 major categories, rather than being listed alphabetically. It is my hope that this reorganization will allow educators and parents to better instruct in units by doing sets of similar projects. While some projects could clearly be placed in multiple categories, the most important category was used when sorting.

Please note that other volumes in the series could still be used to add on to the learning opportunities included in this volume. Make sure to check out:

- 50 Holiday STEM Labs

- 50 More Holiday STEM Labs

- 50 STEM Labs Cards

- 50 STEM Labs Journals

- 50 Weeks of STEM Labs

CATEGORICAL MISSION LISTING - 1

Architecture

1. Cantilever Catastrophe I - Paper
2. Cantilever Catastrophe II - Plastic Straws
3. Cantilever Catastrophe III - Pipe Cleaners
4. Cantilever Catastrophe IV - Wood
5. Cantilever Catastrophe IV - Foil
6. Golden Arches I - Paper
7. Golden Arches II - Plastic Straws
8. Golden Arches III - Foil
9. Golden Arches IV - Pipe Cleaners
10. Golden Arches V - Wood
11. Golden Arches VI - Clay
12. Golden Arches VII - Coins
13. Volume Up

Boats

1. Bayou Fanboats
2. Glass Bottom Boats
3. H.M.S. Speedboat
4. Lead Sinker
5. Old Swimming Raft
6. Straw Rafts
7. What Floats Your Boat I - Clay
8. What Floats Your Boat II - Foil
9. What Floats Your Boat III - Wood

Bridges

1. Bridge to Nowhere
2. Bridging the River Sticks
3. Clean Sweep
4. Gummy Bridges
5. Jet-Puff Bridges
6. Plastic Bridges
7. Rickety Old Bridge
8. Salty Bridges
9. Spaghetti Bridges
10. Stringy Situation
11. Suspension Bridges
12. Tin Highway

CATEGORICAL MISSION LISTING - 2

Cars

1. All Ramped Up
2. Crash Test Dummies
3. Get It Running
4. Marshmallow Snowman Pileup
5. Monster Truck Rally
6. Off to the Races
7. Wooden Cars

Eggs

1. Egg Survivor I - High Falls
2. Egg Survivor II - Smallest Winner
3. Egg Survivor III - Rolling Eggs
4. Egg Survivor IV - Log Flume
5. Egg Survivor V - Mars Lander
6. Egg Survivor VI - Flak Jackets

Flight

1. 10 Seconds and Counting
2. A Breath of Fresh Air
3. Ben Franklin's Kites
4. Come Fly With Me
5. Frisbees
6. Hang Ten
7. Kiteastrophe!
8. Landing Strip
9. Lay it All on the Line
10. Look Out Below!
11. Sticky Planes
12. To the Moon and Back
13. We Can Fly Anything!
14. Whirly Birds

CATEGORICAL MISSION LISTING - 3

Machines

1. Block and Tackle
2. Construction Chaos I - Excavator
3. Construction Chaos II - Crane
4. Construction Chaos III - Forklift
5. Flippers
6. Flushers
7. Gone Fishin'
8. Hand Mixer
9. Marshmallow Mayhem
10. Merry-Go-Round!
11. Pinball Fever
12. Scavenger Bikes
13. Splashing Around
14. Take to the Winds
15. Wheel of Fortune

Strength Test

1. All Tangled Up
2. Can Crusher
3. Daisy Chains
4. Dead Lift
5. Glue-Ten Free
6. Huff and Puff Your House Down
7. Marshmallow Mayhem
8. On a Strong Note...
9. Pipe Dreams
10. Pump You Up
11. Rubber Match
12. Strong as Aluminum
13. Tear the Trampoline I - Plastic Wrap
14. Tear the Trampoline II - Wax Paper
15. Tear the Trampoline III - Tissue
16. Tear the Trampoline IV - Paper
17. Weakest Link
18. Webbed Up

Task Completion

1. Cable Cars I - Weight
2. Cable Cars II - Water
3. Coin Collection
4. Eggs in a Basket
5. Homemade Orchestra I - Percussion
6. Homemade Orchestra II - Strings
7. Homemade Orchestra III - Wind
8. Homemade Orchestra IV - Full Orchestra
9. Let's Get Cooking
10. Lost Your Marbles
11. Pin the Tail on the Balloon
12. Nice Cold Drink
13. Running Uphill
14. Versus I - The Takedown
15. Versus II - The Splashdown
16. Versus III - The Crashdown
17. Versus IV - The Rundown
18. Washboard
19. Water Delivery
20. What's For Breakfast?

CATEGORICAL MISSION LISTING - 4

Throwers

1. 3-2-1 Launch
2. Blowguns
3. Bottle Blasters
4. Crossbows
5. Discus
6. Fire Away!
7. Make it Rain!
8. Marshmallows Away!
9. Marshmallow Blaster
10. Ping Pong Mortar
11. Rain Marbles Down on Them!
12. Target Practice
13. Throwing Money Away
14. Trebuchets

Towers

1. Claymore Tower
2. Foilty Towers
3. Gummy Towers
4. House of Cards
5. High Clips
6. King of Cups
7. Leaning Tower of Pasta
8. Leaning Towers
9. Peanut Tower
10. Pipeworks
11. Starchy Goodness
12. Toothpick Towers
13. Tube Frame Towers

Tracks

1. Copper Road
2. Down the Chute
3. Irrigator
4. Marble Madness
5. Ping Pong Madness
6. Ski Jump
7. Take the High Road
8. Tubular Balls
9. Wooden Railway

150 STEM LABS

Each of the **150 STEM LABS** has the following:

- A snappy **Title**

- A **Brief Description** of the task to be completed

- General **Mission Rules**, suggestions, limitations, and requirements of the task

- **Grading Rubrics** for a Quiz and a Test Grade

- A small **note space** for any changes or adaptations required

- **Category Tags** at the bottom to help you find similar projects

150 STEM Labs

SCIENCE * TECHNOLOGY * ENGINEERING * MATH

Architecture Projects

MISSION: Cantilever Catastrophe I - Paper

BRIEF: You and your team have been selected to make the longest cantilever possible from just tape and paper.

MISSION RULES:

1. Your cantilever must between 2 and 6 inches wide, and 2 to 12 inches tall, but as long as possible. If you are outside these height and width measurements by more than 1/2 inch, you will be penalized.

2. You will work with one or two partners. Teams may not be of more than 3 people.

3. You must only use paper and glue, and tape for your project.

4. The cantilever may be taped to the desk or table on one end.

5. A beam of sorts will project away from the table horizontally as far as possible. The beam distance will only be counted as far as it goes out from the table edge, provided it does not drop more than 2 inches below the table's surface. The distance will only count up to the point in the beam before it descends the acceptable level.

TEACHER'S NOTES: For a more difficult task, add a small weight to the end of the beam, like a penny or large paper clip.

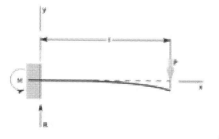

QUIZ GRADE:

A research paper on cantilevers.

- 2-4 pictures of cantilevers 25%

- A labeled sketch and detailed concept idea based on your cantilevers pictures 50%

- Conclusions and reflections based on your results 25%

TEST GRADE:

Your completed design and the results of the test.

- Project Completed = 50%

- 50% of your grade depends on how long your project is compared to the other group projects. The projects that are longest will get more points. Top scores get +50%, and those following get +40%, +30%, or +20%.

- NOTE: There is a -5% penalty for every 1/2 inch your project is out of the specifications.

NOTES:

CATEGORIES: Cantilevers, Length, Materials Strength, Paper

MISSION: Cantilever Catastrophe II - Plastic Straws

BRIEF: You and your team have been selected to make the longest cantilever possible from just tape and plastic straws.

MISSION RULES:

1. Your cantilever must between 2 and 6 inches wide, and 2 to 12 inches tall, but as long as possible. If you are outside these height and width measurements by more than 1/2 inch, you will be penalized.

2. You will work with one or two partners. Teams may not be of more than 3 people.

3. You must only use plastic straws and tape for your project.

4. The cantilever may be taped to the desk or table on one end.

5. A beam of sorts will project away from the table horizontally as far as possible. The beam distance will only be counted as far as it goes out from the table edge, provided it does not drop more than 2 inches below the table's surface. The distance will only count up to the point in the beam before it descends the acceptable level.

TEACHER'S NOTES: For a more difficult task, add a small weight to the end of the beam, like a penny or large paper clip.

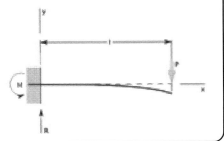

QUIZ GRADE:

A research paper on cantilevers.

- 2-4 pictures of cantilevers 25%

- A labeled sketch and detailed concept idea based on your cantilevers pictures 50%

- Conclusions and reflections based on your results 25%

TEST GRADE:

Your completed design and the results of the test.

- Project Completed = 50%

- 50% of your grade depends on how long your project is compared to the other group projects. The projects that are longest will get more points. Top scores get +50%, and those following get +40%, +30%, or +20%.

- NOTE: There is a -5% penalty for every 1/2 inch your project is out of the specifications.

NOTES:

CATEGORIES: Cantilevers, Length, Materials Strength, Plastic Straws

MISSION: Cantilever Catastrophe III - Pipe Cleaners

BRIEF: You and your team have been selected to make the longest cantilever possible from just tape and plastic straws.

MISSION RULES:

1. Your cantilever must between 2 and 6 inches wide, and 2 to 12 inches tall, but as long as possible. If you are outside these height and width measurements by more than 1/2 inch, you will be penalized.

2. You will work with one or two partners. Teams may not be of more than 3 people.

3. You must only use pipe cleaners and tape for your project. Your teacher will determine how many materials you can use.

4. The cantilever may be taped to the desk or table on one end.

5. A beam of sorts will project away from the table horizontally as far as possible. The beam distance will only be counted as far as it goes out from the table edge, provided it does not drop more than 2 inches below the table's surface. The distance will only count up to the point in the beam before it descends the acceptable level.

TEACHER'S NOTES: For a more difficult task, add a small weight to the end of the beam, like a penny or large paper clip.

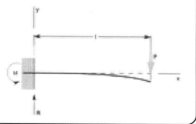

QUIZ GRADE:

A research paper on cantilevers.

- 2-4 pictures of cantilevers 25%

- A labeled sketch and detailed concept idea based on your cantilevers pictures 50%

- Conclusions and reflections based on your results 25%

TEST GRADE:

Your completed design and the results of the test.

- Project Completed = 50%

- 50% of your grade depends on how long your project is compared to the other group projects. The projects that are longest will get more points. Top scores get +50%, and those following get +40%, +30%, or +20%.

- NOTE: There is a -5% penalty for every 1/2 inch your project is out of the specifications.

NOTES:

CATEGORIES: Cantilevers, Length, Materials Strength, Pipe Cleaners

MISSION: Cantilever Catastrophe IV - Wood

BRIEF: You and your team have been selected to make the longest cantilever possible from just toothpicks or popsicle sticks, tape, and glue.

MISSION RULES:

1. Your cantilever must between 2 and 6 inches wide, and 2 to 12 inches tall, but as long as possible. If you are outside these height and width measurements by more than 1/2 inch, you will be penalized.

2. You will work with one or two partners. Teams may not be of more than 3 people.

3. You must only use toothpicks or popsicle sticks for your project along with tape or glue. Your teacher will determine how many materials you can use.

4. The cantilever may be taped to the desk or table on one end.

5. A beam of sorts will project away from the table horizontally as far as possible. The beam distance will only be counted as far as it goes out from the table edge, provided it does not drop more than 2 inches below the table's surface. The distance will only count up to the point in the beam before it descends the acceptable level.

TEACHER'S NOTES: For a more difficult task, add a small weight to the end of the beam, like a penny or large paper clip.

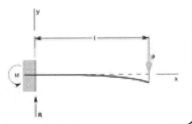

QUIZ GRADE:

A research paper on cantilevers.

- 2-4 pictures of cantilevers 25%

- A labeled sketch and detailed concept idea based on your cantilevers pictures 50%

- Conclusions and reflections based on your results 25%

TEST GRADE:

Your completed design and the results of the test.

- Project Completed = 50%

- 50% of your grade depends on how long your project is compared to the other group projects. The projects that are longest will get more points. Top scores get +50%, and those following get +40%, +30%, or +20%.

- NOTE: There is a -5% penalty for every 1/2 inch your project is out of the specifications.

NOTES:

CATEGORIES: Cantilevers, Length, Materials Strength, Popsicle Sticks, Toothpicks

MISSION: Cantilever Catastrophe V - Foil

BRIEF: You and your team have been selected to make the longest cantilever possible from just foil and tape.

MISSION RULES:

1. Your cantilever must between 2 and 6 inches wide, and 2 to 12 inches tall, but as long as possible. If you are outside these height and width measurements by more than 1/2 inch, you will be penalized.

2. You will work with one or two partners. Teams may not be of more than 3 people.

3. You must only use foil for your project. Your teacher will determine how many materials you can use.

4. The cantilever may be taped to the desk or table on one end.

5. A beam of sorts will project away from the table horizontally as far as possible. The beam distance will only be counted as far as it goes out from the table edge, provided it does not drop more than 2 inches below the table's surface. The distance will only count up to the point in the beam before it descends the acceptable level.

TEACHER'S NOTES: For a more difficult task, add a small weight to the end of the beam, like a penny or large paper clip.

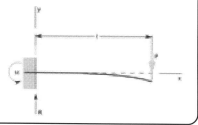

QUIZ GRADE:

A research paper on cantilevers.

- 2-4 pictures of cantilevers 25%

- A labeled sketch and detailed concept idea based on your cantilevers pictures 50%

- Conclusions and reflections based on your results 25%

TEST GRADE:

Your completed design and the results of the test.

- Project Completed = 50%

- 50% of your grade depends on how long your project is compared to the other group projects. The projects that are longest will get more points. Top scores get +50%, and those following get +40%, +30%, or +20%.

- NOTE: There is a -5% penalty for every 1/2 inch your project is out of the specifications.

NOTES:

CATEGORIES: Cantilevers, Foil, Length, Materials Strength

MISSION: Golden Arches I - Paper

BRIEF:

You and your team have been selected to design an arch that is as tall and wide as possible.

MISSION RULES:

1. You will design an arch that is as tall and as wide as possible.

2. Your finished project must be built of only paper and tape.

3. Your project must be completely freestanding and may not be attached to the floor or a table surface.

4. Your teacher will determine your materials limit for the arch.

5. You will work with one or two partners. Teams may be of no more than 3 people.

6. While a perfect curve is hard to attain, some semblance of an arch must be created. A simple V or single-pitched roof is not acceptable.

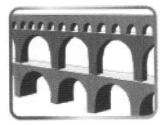

QUIZ GRADE:

Create a blueprint design for your ideas

- Sketch 25%

- Sketch is labeled 25%

- Explanation of strategies 25%

- Conclusions and reflections based on your results 25%

TEST GRADE:

Your completed design and the results of the test.

- Project Completed = 50%

- 50% of your project's score depends on the height and width of your project as compared to other projects.

- A formula of width X height will be used. Width is at the widest part near the base, and height is as the peak of the arch.

- NOTE: The arch with the highest score will automatically get 100%

NOTES:

CATEGORIES: Arches, Height, Paper

MISSION: Golden Arches II - Plastic Straws

BRIEF: You and your team have been selected to design an arch that is as tall and wide as possible.

MISSION RULES:

1. You will design an arch that is as tall and as wide as possible.

2. Your finished project must be built of only plastic straws, tape, and a small amount of card stock.

3. Your project must be completely freestanding and may not be attached to the floor or a table surface.

4. Your teacher will determine your materials limit for the arch.

5. You will work with one or two partners. Teams may be of no more than 3 people.

6. While a perfect curve is hard to attain, some semblance of an arch must be created. A simple V or single-pitched roof is not acceptable.

QUIZ GRADE:

Create a blueprint design for your ideas

- Sketch 25%

- Sketch is labeled 25%

- Explanation of strategies 25%

- Conclusions and reflections based on your results 25%

TEST GRADE:

Your completed design and the results of the test.

- Project Completed = 50%

- 50% of your project's score depends on the height and width of your project as compared to other projects.

- A formula of width X height will be used. Width is at the widest part near the base, and height is as the peak of the arch.

- NOTE: The arch with the highest score will automatically get 100%

NOTES:

CATEGORIES: Arches, Height, Plastic Straws

MISSION: Golden Arches III - Foil

BRIEF: You and your team have been selected to design an arch that is as tall and wide as possible.

MISSION RULES:

1. You will design an arch that is as tall and as wide as possible.

2. Your finished project must be built of only foil and a small amount of tape.

3. Your project must be completely freestanding and may not be attached to the floor or a table surface.

4. Your teacher will determine your materials limit for the arch.

5. You will work with one or two partners. Teams may be of no more than 3 people.

6. While a perfect curve is hard to attain, some semblance of an arch must be created. A simple V or single-pitched roof is not acceptable.

QUIZ GRADE:

Create a blueprint design for your ideas

- Sketch 25%

- Sketch is labeled 25%

- Explanation of strategies 25%

- Conclusions and reflections based on your results 25%

TEST GRADE:

Your completed design and the results of the test.

- Project Completed = 50%

- 50% of your project's score depends on the height and width of your project as compared to other projects.

- A formula of width X height will be used. Width is at the widest part near the base, and height is as the peak of the arch.

- NOTE: The arch with the highest score will automatically get 100%

NOTES:

CATEGORIES: Arches, Foil, Height

MISSION: Golden Arches IV - Pipe Cleaners

BRIEF:

You and your team have been selected to design an arch that is as tall and wide as possible.

MISSION RULES:

1. You will design an arch that is as tall and as wide as possible.

2. Your finished project must be built of only pipe cleaners.

3. Your project must be completely freestanding and may not be attached to the floor or a table surface.

4. Your teacher will determine your materials limit for the arch.

5. You will work with one or two partners. Teams may be of no more than 3 people.

6. While a perfect curve is hard to attain, some semblance of an arch must be created. A simple V or single-pitched roof is not acceptable.

QUIZ GRADE:

Create a blueprint design for your ideas

- Sketch 25%

- Sketch is labeled 25%

- Explanation of strategies 25%

- Conclusions and reflections based on your results 25%

TEST GRADE:

Your completed design and the results of the test.

- Project Completed = 50%

- 50% of your project's score depends on the height and width of your project as compared to other projects.

- A formula of width X height will be used. Width is at the widest part near the base, and height is as the peak of the arch.

- NOTE: The arch with the highest score will automatically get 100%

NOTES:

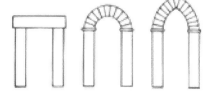

CATEGORIES: Arches, Height, Pipe Cleaners

MISSION: Golden Arches V - Wood

BRIEF: You and your team have been selected to design an arch that is as tall and wide as possible.

MISSION RULES:

1. You will design an arch that is as tall and as wide as possible.

2. Your finished project must be built of only toothpicks and/or popsicle sticks and glue.

3. Your project must be completely freestanding and may not be attached to the floor or a table surface.

4. Your teacher will determine your materials limit for the arch.

5. You will work with one or two partners. Teams may be of no more than 3 people.

6. While a perfect curve is hard to attain, some semblance of an arch must be created. A simple V or single-pitched roof is not acceptable.

QUIZ GRADE:

Create a blueprint design for your ideas

- Sketch 25%

- Sketch is labeled 25%

- Explanation of strategies 25%

- Conclusions and reflections based on your results 25%

TEST GRADE:

Your completed design and the results of the test.

- Project Completed = 50%

- 50% of your project's score depends on the height and width of your project as compared to other projects.

- A formula of width X height will be used. Width is at the widest part near the base, and height is as the peak of the arch.

- NOTE: The arch with the highest score will automatically get 100%

NOTES:

CATEGORIES: Arches, Height, Popsicle Sticks, Toothpicks

MISSION: Golden Arches VI - Clay

BRIEF:

You and your team have been selected to design an arch that is as tall and wide as possible.

MISSION RULES:

1. You will design an arch that is as tall and as wide as possible.

2. Your finished project must be built of only clay.

3. Your project must be completely freestanding and may not be attached to the floor or a table surface.

4. Your teacher will determine your materials limit for the arch.

5. You will work with one or two partners. Teams may be of no more than 3 people.

6. While a perfect curve is hard to attain, some semblance of an arch must be created. A simple V or single-pitched roof is not acceptable.

QUIZ GRADE:

Create a blueprint design for your ideas

- Sketch 25%

- Sketch is labeled 25%

- Explanation of strategies 25%

- Conclusions and reflections based on your results 25%

TEST GRADE:

Your completed design and the results of the test.

- Project Completed = 50%

- 50% of your project's score depends on the height and width of your project as compared to other projects.

- A formula of width X height will be used. Width is at the widest part near the base, and height is as the peak of the arch.

- NOTE: The arch with the highest score will automatically get 100%

NOTES:

CATEGORIES: Arches, Clay, Height

MISSION: Golden Arches VII - Coins

BRIEF:
You and your team have been selected to design an arch that is as tall and wide as possible from just coins.

MISSION RULES:

1. You will design an arch that is as tall and as wide as possible.

2. Your finished project must be built of only coins.

3. Your project must be completely freestanding and may not be attached to the floor or a table surface.

4. Your teacher will determine your materials limit for the arch.

5. You will work with one or two partners. Teams may be of no more than 3 people.

6. While a perfect curve is hard to attain, some semblance of an arch must be created. A simple V or single-pitched roof is not acceptable.

QUIZ GRADE:

Create a blueprint design for your ideas

- Sketch 25%

- Sketch is labeled 25%

- Explanation of strategies 25%

- Conclusions and reflections based on your results 25%

TEST GRADE:

Your completed design and the results of the test.

- Project Completed = 50%

- 50% of your project's score depends on the height and width of your project as compared to other projects.

- A formula of width X height will be used. Width is at the widest part near the base, and height is as the peak of the arch.

- NOTE: The arch with the highest score will automatically get 100%

NOTES:

CATEGORIES: Arches, Coins, Height

MISSION: Volume Up

BRIEF: You and your team have been selected to the largest possible structure possible with only popsicle sticks and rubber bands.

MISSION RULES:

1. Your device may be any possible dimension provided it is made with only the provided amount of rubber bands and popsicle sticks.

2. You will work with one or two partners. Teams may not be of more than 3 people.

3. You must only use rubber bands and popsicle sticks for your project. You teacher will determine how many materials you may use.

4. If the project is roughly rectangular in shape, length x width x height will be used to determine volume, rounding to the nearest inch before multiplying.

5. If the project is more spherical in shape, the formula 4/3 pi r cubed should be used.

QUIZ GRADE:

A blueprint design of your idea

- Sketch 25%

- Sketch is labeled 25%

- Explanation of strategy 25%

- Conclusions and reflections based on your results 25%

TEST GRADE:

Your completed design and the results of the test.

- Project Completed = 50%

- 50% of your grade depends on how large (volume) your project is compared to others.

- The biggest project gets an automatic 100%

NOTES:

CATEGORIES: Popsicle Sticks, Rubber Bands, Volume

150 STEM Labs

SCIENCE * TECHNOLOGY * ENGINEERING * MATH

Boats Projects

MISSION: Bayou Fanboats

BRIEF: You and your team have been selected to make balloon-powered boat that goes as fast and as far as possible.

MISSION RULES:

1. You will design a boat that is balloon-powered and attached to a string. The rocket ship must slide along the string to cross the body of water.

2. Your boat must be built from a single balloon, a straw, tape or glue, notecards, and other teacher-approved materials.

3. You will work with a single partner. Teams may not be of more than 2 people.

4. The straw will be used to slide along the line that is strung horizontally above a body of water. Both ends will be secured during tests. Inflated balloons will be attached to the balloon boat. Letting air from the balloons should propel the project across the water as fast as possible.

5. If your boat does not clear the distance with one use of the balloon, your teacher has the option of measuring the distance traveled or keeping the clock running while you refill the balloon as many times as needed to clear the course.

TEACHER'S NOTES: You will need some sort of body of water. A set of 1-3 rain gutters coupled together with caps at the end will hold water. Then you can run the string/fishing line parallel to the surface of the water.

QUIZ GRADE:

Create a blueprint designs for your ideas

- Sketch 25%

- Sketch is labeled 25%

- Explanation of strategies 25%

- Conclusions and reflections based on your results 25%

TEST GRADE:

Your completed design and the results of the test.

- Project Completed = 50%

50% of your grade depends on how fast your project is compared to others.

- Other places = +10-20%

- Third Place = +30%

- Second Place = +40%

- Fastest Boat = +50%

NOTES:

CATEGORIES: Balloons, Boats, Buoyancy, Distance, Speed, Water

MISSION: Glass Bottom Boats

BRIEF: You and your team have been selected to make a boat from pipe cleaners and plastic wrap to help your marshmallow family cross a lake!

MISSION RULES:

1. You will design a boat that built entirely of only plastic wrap and pipe cleaners.

2. Your boat must have an area to house a family of four mini marshmallow family members (built from 2-3 mini marshmallows and toothpicks).

3. You will work with a single partner. Teams may not be of more than 2 people.

4. The boat must have at least one dimension greater than 6 inches.

5. The boat will be pushed across the 'lake' or tub of water with a fan, so you may want to develop some sort of sail or wind catcher.

6. Your boat must not tip over or get the family wet!

TEACHER'S NOTES: You will need some sort of body of water. A large sink, a water tub, or even a giant trash can filled with water could work. A rain set of 1-3 rain gutters coupled together with caps at the end may hold water, too.

You also need a small fan to blow the boats across the water.

QUIZ GRADE:

Create a blueprint designs for your ideas

- Sketch 25%

- Sketch is labeled 25%

- Explanation of strategies 25%

- Conclusions and reflections based on your results 25%

TEST GRADE:

Your completed design and the results of the test.

- Project Completed = 40%

60% of your grade depends on if your family members stay dry. Each one is worth 15%.

NOTE: You may be required to make multiple runs across the lake, too!

NOTES:

CATEGORIES: Boats, Marshmallows, Pipe Cleaners, Plastic Wrap, Water, Wind

MISSION: H.M.S. Speedboat

BRIEF: You and your team have been selected to design the fastest wind-powered land boat possible.

MISSION RULES:

1. You will design and build a sailboat from notecards, glue, tape, toothpicks, popsicle sticks, rubber bands, straws, or other scavenged household materials.

2. Your finished boat must roll or move across land on wind power alone.

3. You will work with one or two partners. Teams may be of no more than 3 people.

4. Your objective is to make the fastest boat possible, powered by natural wind or by a fan.

QUIZ GRADE:

Create a blueprint designs for your ideas

- Sketch 25%

- Sketch is labeled 25%

- Explanation of strategies 25%

- Conclusions and reflections based on your results 25%

TEST GRADE:

Your completed design and the results of the test.

- Project Completed = 50%

50% of your grade depends on how fast your project is compared to others.

- Other places = +20%

- Third Place = +30%

- Second Place = +40%

- Fastest Boat = +50%

NOTES:

CATEGORIES: Boats, Cars, Scavengers, Speed, Wind

MISSION: Lead Sinker

BRIEF: You and your team have been selected to design a device that can sink a ping pong ball and to make it as small as possible.

MISSION RULES:

1. You will design a device from scavenged materials. Materials must be approved by the teacher.

2. The device must sink the ping pong ball and stay sunk for at least 10 seconds.

3. You will work with one or two partners. Teams may be of no more than 3 people.

4. You may get test trials to see if your device works. The number of test runs depends on your teacher.

5. Your device must have a cup or receiver of some sort to hold the ping pong ball. The ping pong ball must be at least partially visible at all times.

6. Your device may not have any dimension larger than 6 inches.

TEACHER'S NOTES: You may want to disallow very heavy materials. If this seems too easy, decrease the dimensions allowed or increase the number of ping pong balls.

QUIZ GRADE:

A blueprint design of your idea

- Sketch 25%

- Sketch is labeled 25%

- Explanation of strategy 25%

- Conclusions and reflections based on your results 25%

TEST GRADE:

Your completed design and the results of the test.

- Project Completed = 50% for successfully sinking a ping pong ball.

- 50% of your grade depends on how small your project is compared to the others. You will be awarded these points in 10% increments.

- OPTIONAL: Points may be awarded differently for tests requiring you to sink more than one ping pong ball.

NOTES:

CATEGORIES: Boats, Buoyancy, Ping Pong Balls, Scavengers, Water, Weight

MISSION: The Old Swimming Raft

BRIEF:

You and your team have been selected to design a ping pong ball raft that can hold the most paperclips without sinking.

MISSION RULES:

1. You will design a raft with 4 ping pong balls and scavenged materials like: paper, toothpicks, tape, cardboard, plastic...

2. The boat must float. You will be given 3 chances to see if it floats prior to the real testing.

3. You will work with one or two partners. Teams may be of no more than 3 people.

4. After putting your floating boat into the water for the actual test, paperclips will be added until it sinks or tips over and dumps the paperclips.

5. Your boat must have a cup or receiver of some sort to hold the weights. If this fills up and there is no more room to add weight, no more weight will be added.

TEACHER'S NOTE: Pennies, measured amounts of sand, or graduated weights can also be used instead of paperclips.

QUIZ GRADE:

A blueprint design of your idea

- Sketch 25%

- Sketch is labeled 25%

- Explanation of strategy 25%

- Conclusions and reflections based on your results 25%

TEST GRADE:

Your completed design and the results of the test.

- Project Completed = 50% for successfully floating.

- 50% of your grade depends on how much weight the boat holds. The more it holds in comparison to the other teams, the better you do.

- You will be awarded these points in 10% increments. The boat that holds the most clips automatically gets 100%

NOTES:

CATEGORIES: Boats, Buoyancy, Ping Pong Balls, Scavengers, Water, Weight

MISSION: Straw Rafts

BRIEF: You and your team have been selected to design a raft from plastic straws that can hold the most weight without sinking.

MISSION RULES:

1. You will design a raft with only 10 plastic straws and glue or tape.

2. The raft must float. You will be given 3 chances to see if it floats prior to the real testing.

3. You will work with one or two partners. Teams may be of no more than 3 people.

4. Your boat must have a spot to place a cup or receiver of some sort to hold the weights.

5. After putting your floating boat into the water for the actual test, paperclips will be added until it sinks or tips over and dumps the paperclips.

TEACHER'S NOTE: Pennies, measured amounts of sand, or graduated weights can also be used instead of paperclips.

QUIZ GRADE:

A blueprint design of your idea

- Sketch 25%

- Sketch is labeled 25%

- Explanation of strategy 25%

- Conclusions and reflections based on your results 25%

TEST GRADE:

Your completed design and the results of the test.

- Project Completed = 50% for successfully floating.

- 50% of your grade depends on how much weight the boat holds. The more it holds in comparison to the other teams, the better you do.

- You will be awarded these points in 10% increments. The boat that holds the most clips automatically gets 100%

NOTES:

CATEGORIES: Boats, Buoyancy, Plastic Straws, Water, Weight

MISSION: What Floats Your Boat I - Clay

BRIEF:
You and your team have been selected to design a clay boat that can hold the most paperclips without sinking.

MISSION RULES:

1. You will design a boat from a block of clay to be placed in water.

2. The boat must float. You will be given 3 chances to see if it floats prior to the real testing. It is recommended that you towel it off afterward, to keep your clay from getting too wet and slimy.

3. You will work with one or two partners. Teams may be of no more than 3 people.

4. After putting your floating boat into the water for the actual test, paperclips will be added until it sinks.

QUIZ GRADE:

A blueprint design of your idea

- Sketch 25%

- Sketch is labeled 25%

- Explanation of strategy 25%

- Conclusions and reflections based on your results 25%

TEST GRADE:

Your completed design and the results of the test.

- Project Completed = 50% for successfully floating.

- 50% of your grade depends on how much weight the boat holds. The more it holds in comparison to the other teams, the better you do.

- You will be awarded these points in 10% increments. The boat that holds the most clips automatically gets 100%

NOTES:

CATEGORIES: Boats, Buoyancy, Clay, Water

MISSION: What Floats Your Boat II - Foil

BRIEF: You and your team have been selected to design a foil boat that can hold the most paperclips without sinking.

MISSION RULES:

1. You will design a boat from a 6"x6" piece of foil to be placed in water.

2. The boat must float. You will be given 3 chances to see if it floats prior to the real testing.

3. You will work with one or two partners. Teams may be of no more than 3 people.

4. After putting your floating boat into the water for the actual test, paperclips will be added until it sinks.

QUIZ GRADE:

A blueprint design of your idea

- Sketch 25%

- Sketch is labeled 25%

- Explanation of strategy 25%

- Conclusions and reflections based on your results 25%

TEST GRADE:

Your completed design and the results of the test.

- Project Completed = 50% for successfully floating.

- 50% of your grade depends on how much weight the boat holds. The more it holds in comparison to the other teams, the better you do.

- You will be awarded these points in 10% increments. The boat that holds the most clips automatically gets 100%

NOTES:

CATEGORIES: Boats, Buoyancy, Foil, Water

MISSION: What Floats Your Boat III - Wood

BRIEF: You and your team have been selected to design a wooden boat that can hold the most paperclips without sinking.

MISSION RULES:

1. You will design a boat from popsicle sticks, toothpicks, 2 3x5 index notecards, and glue.

2. The boat may be no longer in any one dimension than one popsicle stick.

3. The boat must float. You will be given 3 chances to see if it floats prior to the real testing.

4. You will work with one or two partners. Teams may be of no more than 3 people.

5. After putting your floating boat into the water for the actual test, paperclips will be added in increments of 10, 25, 50, or 100 until it sinks.

QUIZ GRADE:

A blueprint design of your idea

- Sketch 25%

- Sketch is labeled 25%

- Explanation of strategy 25%

- Conclusions and reflections based on your results 25%

TEST GRADE:

Your completed design and the results of the test.

- Project Completed = 50% for successfully floating.

- 50% of your grade depends on how much weight the boat holds. The more it holds in comparison to the other teams, the better you do.

- You will be awarded these points in 10% increments. The boat that holds the most clips automatically gets 100%

NOTES:

CATEGORIES: Boats, Buoyancy, Toothpicks, Water

150 STEM Labs

SCIENCE * TECHNOLOGY * ENGINEERING * MATH

Bridges Projects

MISSION: Bridge to Nowhere

BRIEF: You and your team have been selected to make as strong of a bridge as possible using only paper and glue.

MISSION RULES:

1. You will research bridges and get ideas for a concept for your bridge design.

2. Your bridge must be 36 inches long, between 2 and 6 inches wide, and 2 to 12 inches tall. If you are outside these measurements by more than 1/2 inch, you will be penalized.

3. You will work with a single partner. Teams may not be of more than 2 people.

4. You must only use paper and glue for your project.

5. The bridge must have a place in the center of it where a tray can be attached to hold weight. Projects that hold more weight score better.

QUIZ GRADE:

A research paper on bridges.

- 2-4 pictures of bridges 25%

- A concept idea based on your bridge pictures 50%

- Conclusions and reflections based on your results 25%

TEST GRADE:

Your completed design and the results of the test.

- Project Completed = 50%

- 50% of your grade depends on how much weight your project holds compared to the other group's projects. The projects that do best will get more points.

- NOTE: There is a -5% penalty for every 1/2 inch your project is out of the specifications.

NOTES:

CATEGORIES: Bridges, Materials Strength, Paper, Weight

MISSION: Bridging the River Sticks

BRIEF: You and your team have been selected to make the longest bridge possible from just toothpicks and glue.

MISSION RULES:

1. You will research bridges and get ideas for a concept for your bridge design. Pay special attention to the geometry involved in bridge construction.

2. Your bridge must be as long as possible, between 2 and 6 inches wide, and 2 to 12 inches tall. If you are out- side these measurements by more than 1/2 inch, you will be penalized 5% per 1/2 inch.

3. You will work with one to three partners. Teams may not be of more than 4 people.

4. You may use only glue and toothpicks in the final project, as well as wax paper as a working surface to help things not stick as they dry. (there may be some bracing used when the project's glue is drying, but they must not be present in the finished project).

5. Longer projects will score better.

QUIZ GRADE:

A research paper on bridges.

- 2-4 pictures of bridges 25%

- A concept idea based on your bridge pictures 50%

- Conclusions and reflections based on your results 25%

TEST GRADE:

Your completed design and the results of the test.

- Project Completed = 50%

- 50% of your grade depends on how long your project is compared to the other group's projects. The projects that do best will get more points.

- NOTE: There is a -5% penalty for every 1/2 inch your project is out of the specifications.

NOTES:

CATEGORIES: Bridges, Materials Strength, Toothpicks, Weight

MISSION: Clean Sweep

BRIEF: You and your team have been selected to make as strong of a bridge as possible using only pipe cleaners.

MISSION RULES:

1. You will research bridges and get ideas for a concept for your bridge design.

2. Your bridge must be 18 inches long, between 2 and 6 inches wide, and 2 to 12 inches tall. If you are outside these measurements by more than 1/2 inch, you will be penalized.

3. You will work with two or three partners. Teams may not be of more than 4 people.

4. You must only use pipe cleaners for your project. You teacher will determine how many you get.

5. The bridge must have a place in the center of it where a tray can be attached to hold weight. Projects that hold more weight score better.

QUIZ GRADE:

A research paper on bridges.

- 2-4 pictures of bridges 25%

- A concept idea based on your bridge pictures 50%

- Conclusions and reflections based on your results 25%

TEST GRADE:

Your completed design and the results of the test.

- Project Completed = 50%

- 50% of your grade depends on how much weight your project holds compared to the other group's projects. The projects that do best will get more points.

- NOTE: There is a -5% penalty for every 1/2 inch your project is out of the specifications.

NOTES:

CATEGORIES: Bridges, Materials Strength, Pipe Cleaners, Weight

MISSION: Gummy Bridges

BRIEF: You and your team have been selected to make as long of a bridge as possible using only gumdrops and toothpicks.

MISSION RULES:

1. You will research bridges and get ideas for a concept for your bridge design.

2. Your bridge must be as long as possible using the provided materials.

3. You will work with one or two partners. Teams may not be of more than 3 people.

4. You must only use gumdrops and toothpicks for your project. You teacher will determine how many materials you may use.

5. The bridge must not collapse for at least 10 seconds when it is being tested.

QUIZ GRADE:

A research paper on bridges.

- 2-4 pictures of bridges 25%

- A concept idea based on your bridge pictures 50%

- Conclusions and reflections based on your results 25%

TEST GRADE:

Your completed design and the results of the test.

- Project Completed = 50%

- 50% of your grade depends on how long your project is compared to the other group's projects. The projects that do best will get more points.

NOTES:

CATEGORIES: Bridges, Gumdrops, Length, Toothpicks

MISSION: Jet-Puff Bridges

BRIEF: You and your team have been selected to make as strong of a bridge as possible from pasta and marshmallows!

MISSION RULES:

1. You will research bridges and get ideas for a concept for your bridge design.

2. Your bridge must be 18 inches long, between 2 and 6 inches wide, and 2 to 12 inches tall. If you are outside these measurements by more than 1/2 inch, you will be penalized.

3. You will work with two or three partners. Teams may not be of more than 4 people.

4. You must only use pasta and marshmallows for your project. You teacher will determine how many materials you may use.

5. The bridge must have a place in the center of it where a tray can be attached to hold weight. Projects that hold more weight score better.

QUIZ GRADE:

A research paper on bridges.

- 2-4 pictures of bridges 25%

- A concept idea based on your bridge pictures 50%

- Conclusions and reflections based on your results 25%

TEST GRADE:

Your completed design and the results of the test.

- Project Completed = 50%

- 50% of your grade depends on how much weight your project holds compared to the other group's projects. The projects that do best will get more points.

- NOTE: There is a -5% penalty for every 1/2 inch your project is out of the specifications.

NOTES:

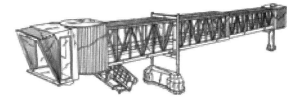

CATEGORIES: Bridges, Marshmallows, Materials Strength, Pasta, Weight

MISSION: Plastic Bridges

BRIEF: You and your team have been selected to make the longest bridge possible from just tape and plastic straws.

MISSION RULES:

1. You will design a device that holds a small container on top. The container will be filled gradually with weight (sand, paperclips, marbles, or coins), and must hold as much weight as possible.

2. Your teacher will determine the maximum number of pipe cleaners you may use in your project. You may use only pipe cleaners for your construction.

3. You will work with one to two partners. Teams may not be of more than 3 people.

4. Your device must be free-standing and movable. It cannot be attached to any surface.

QUIZ GRADE:

A research paper on bridges.

- 2-4 pictures of bridges 25%

- A labeled sketch and de- tailed concept idea based on your bridge pictures 50%

- Conclusions and reflections based on your results 25%

TEST GRADE:

Your completed design and the results of the test.

- Project Completed = 50%

- 50% of your grade depends on how long your project is compared to the other group's projects. The projects that are longest will get more points. Top scores get +50%, and those following get +40%, +30%, or +20%.

- NOTE: There is a -5% penalty for every 1/2 inch your project is out of the specifications.

NOTES:

CATEGORIES: Bridges, Length, Plastic Straws

MISSION: Rickety Old Bridge

BRIEF: You and your team have been selected to make a rope bridge from twine and note cards.

MISSION RULES:

1. You will design a bridge using only twine and notecards for planks. There may be no glue or tape.

2. Your bridge cannot use any piece of twine longer than 6", or a piece of notecard larger than 1/4 of a whole card.

3. Your bridge must be at least 18" long, and 3" wide. There must also be some sort of handrails.

4. You will work with 1 or 2 partners. Teams may not be of more than 3 people.

5. The rope bridge must be able to anchor to a board with 2 posts or some setup your teacher provides to anchor each side of the bridge to.

- TEACHER'S OPTION: You must be able to push a hot wheels/matchbox car across the bridge without it getting stuck.

- TEACHER'S OPTION #2: Test the weight the bridge can hold by putting graduated weights onto the project.

QUIZ GRADE:

Create a blueprint design for your ideas

- Sketch 25%

- Sketch is labeled 25%

- Explanation of strategies 25%

- Conclusions and reflections based on your results 25%

TEST GRADE:

Your completed design and the results of the test.

- Project Completed = 60%

- Project Length: 18"=60%, 21"=70", 24"=80%, 27"=90%, 30"=100%

- NOTE: Other scores may vary according to bonuses and penalties assessed for any of the teacher's options chosen.

NOTES:

CATEGORIES: Bridges, Cars, Length, Materials Strength, String, Weight

MISSION: Salty Bridges

BRIEF: You and your team have been selected to make the longest bridge as possible from pretzel sticks and glue.

MISSION RULES:

1. You will research bridges and get ideas for a concept for your bridge design.

2. Your bridge must be as long as possible. It may be any height or width required to accomplish this.

3. You will work with two or three partners. Teams may not be of more than 4 people.

4. You must only use thin pretzel sticks and glue for your project. You teacher will determine how many materials you may use.

5. The bridge must not be attached to any surface, but must have a place on each side where it can be rested on a surface, so it can set up to span between two tables, chairs, or something along those lines.

TEACHER'S NOTES: Wax paper is advised for a surface on which to dry the glue and pretzel sticks as it is being assembled.

QUIZ GRADE:

A research paper on bridges.

- 2-4 pictures of bridges 25%

- A concept idea based on your bridge pictures 50%

- Conclusions and reflections based on your results 25%

TEST GRADE:

Your completed design and the results of the test.

- Project Completed = 50%

- 50% of your grade depends on how long your project is compared to the other group's projects. The projects that do best will get more points.

NOTES:

CATEGORIES: Bridges, Glue, Length, Pretzels

MISSION: Spaghetti Bridges

BRIEF: You and your team have been selected to make as strong of a bridge as possible from pasta and glue!

MISSION RULES:

1. You will research bridges and get ideas for a concept for your bridge design.

2. Your bridge must be 18 inches long, between 2 and 6 inches wide, and 2 to 12 inches tall. If you are outside these measurements by more than 1/2 inch, you will be penalized.

3. You will work with two or three partners. Teams may not be of more than 4 people.

4. You must only use pasta and glue for your project. You teacher will determine how many materials you may use.

5. The bridge must have a place in the center of it where a tray can be attached to hold weight. Projects that hold more weight score better.

QUIZ GRADE:

A research paper on bridges.

- 2-4 pictures of bridges 25%

- A concept idea based on your bridge pictures 50%

- Conclusions and reflections based on your results 25%

TEST GRADE:

Your completed design and the results of the test.

- Project Completed = 50%

- 50% of your grade depends on how much weight your project holds compared to the other group's projects. The projects that do best will get more points.

- NOTE: There is a -5% penalty for every 1/2 inch your project is out of the specifications.

NOTES:

CATEGORIES: Bridges, Glue, Materials Strength, Pasta, Weight

MISSION: Stringy Situation

BRIEF: You and your team have been selected to make as strong of a bridge as possible from string or yarn and glue!

MISSION RULES:

1. You will research bridges and get ideas for a concept for your bridge design.

2. Your bridge must be 18 inches long, between 2 and 6 inches wide, and 2 to 12 inches tall. If you are outside these measurements by more than 1/2 inch, you will be penalized.

3. You will work with two or three partners. Teams may not be of more than 4 people.

4. You must only use string (or yarn) and glue for your project. Your teacher will determine the amount of materials you receive.

5. The bridge must have a place in the center of it where a tray can be attached to hold weight. Projects that hold more weight score better.

TEACHER'S NOTES: It is suggested that you use wax paper to set up 'cables' of thread or yarn. Glue will help them set into a specific shape when it dries. Then they can be layered and attached to form the bridge.

QUIZ GRADE:

A research paper on bridges.

- 2-4 pictures of bridges 25%

- A concept idea based on your bridge pictures 50%

- Conclusions and reflections based on your results 25%

TEST GRADE:

Your completed design and the results of the test.

- Project Completed = 50%

- 50% of your grade depends on how much weight your project holds compared to the other group's projects. The projects that do best will get more points.

- NOTE: There is a -5% penalty for every 1/2 inch your project is out of the specifications.

NOTES:

CATEGORIES: Bridges, Glue, Materials Strength, String, Weight, Yarn

MISSION: Suspension Bridges

BRIEF: You and your team have been selected to make as strong of a suspension bridge as possible!

MISSION RULES:

1. You will research suspension bridges and get ideas for a concept for your bridge design.

2. Your bridge must be 18 inches long, between 2 and 6 inches wide, and 2 to 12 inches tall. If you are outside these measurements by more than 1/2 inch, you will be penalized.

3. You will work with two or three partners. Teams may not be of more than 4 people.

4. You may use any approved materials from home or school.

5. The bridge must have a string or line running from one side to the other. A cup will be hung from the line in the center of the bridge, where weight will be added.

TEACHER'S NOTES: Paperclips or coins make great weights. The cup will need some sort of hooks, probably paperclips, to attach it to the suspension lines. You may wish to suggest towers/bracing of some sort on the designs.

QUIZ GRADE:

A research paper on suspension bridges.

- 2-4 pictures of bridges 25%

- A concept idea based on your bridge pictures 50%

- Conclusions and reflections based on your results 25%

TEST GRADE:

Your completed design and the results of the test.

- Project Completed = 50%

- 50% of your grade depends on how much weight your project holds compared to the other group's projects. The projects that do best will get more points.

- NOTE: There is a -5% penalty for every 1/2 inch your project is out of the specifications.

NOTES:

CATEGORIES: Bridges, Materials Strength, Scavengers, String, Weight

MISSION: Tin Highway

BRIEF: You and your team have been selected to make the strongest bridge possible from tin foil.

MISSION RULES:

- You will research bridges and get ideas for a concept for your bridge design.

- Your bridge must be 18 inches long, between 2 and 4 inches wide, and 2 to 6 inches tall. If you are outside these measurements by more than 1/2 inch, you will be penalized.

- You will work with a single partner. Teams may not be of more than 2 people.

- You must only use tin foil for your project. Your teacher will determine how much foil you get.

- The bridge must have a place in the center of it where a small tray can be attached to hold weight (probably paperclips or pennies). Projects that hold more weight score better.

QUIZ GRADE:

A research paper on bridges.

- 2-4 pictures of bridges 25%

- A labeled sketch and de- tailed concept idea based on your bridge pictures 50%

- Conclusions and reflections based on your results 25%

TEST GRADE:

Your completed design and the results of the test.

- Project Completed = 50%

- 50% of your grade depends on how much weight your project is compared to the other group's projects. The projects that hold the most will get more points. Top scores get +50%, and those following get +40%, +30%, or +20%.

- NOTE: There is a -5% penalty for every 1/2 inch your project is out of the specifications.

NOTES:

CATEGORIES: Bridges, Foil, Materials Strength, Weight

150 STEM Labs

SCIENCE * TECHNOLOGY * ENGINEERING * MATH

Cars Projects

MISSION: All Ramped Up

BRIEF: You and your team have been selected to design a ramp for a matchbox car to jump as far as possible with.

MISSION RULES:

1. You will design a car ramp from scavenged supplies, including: tape, straws, paper clips, note cards, cardboard tubes/boxes, paper, card stock, and other approved supplies.

2. Your finished ramp must accommodate a car provided by your teacher, allowing the car to roll down and jump off the track.

3. It should be no more than 18 inches high, no longer than 18 inches, and no wider than 6 inches.

4. You will work with one or two partners. Teams may be of no more than 3 people.

5. Success will be measured by how far your car jumps after exiting your ramp.

6. Your teacher will determine the number of trials you receive. If more than 1 trial is done, an average score will be used.

QUIZ GRADE:

- Written step-by-step knot-tying procedures 25%

- Demonstration of knot tying by both partners 25%

- Explanation of strategy 25%

- Conclusions and reflections based on your results 25%

TEST GRADE:

Your completed design and the results of the test.

- Project Completed = 50%

- 50% of your grade depends on how far your car jumps after leaving your ramp.

- The team that builds the ramp that allows the car to jump the farthest will get an automatic 100%. Other teams will receive 10-40%, based on performance.

NOTES:

RAMP
30
MPH

CATEGORIES: Cars, Distance, Tracks

MISSION: Crash Test Dummies

BRIEF: You and your team have been selected to design a stunt car that can protect passengers to the end of the track.

MISSION RULES:

1. You will design a car from scavenged supplies, including: tape, straws, paper clips, note cards, cardboard tubes/boxes, paper, card stock, and other approved supplies.

2. Your finished cars must fit on a track that your teacher has created, likely to include things like bumps, jumps, hills, and a wall to crash into at the very end.

3. You will work with one or two partners. Teams may be of no more than 3 people.

4. Success will be measured by how many of your car's passengers are still in the car at the end of the track.

5. Your project design may not completely close off or secure your passengers to the car itself. Your passengers must be at least partially visible at all times.

QUIZ GRADE:

Create a blueprint design for your ideas

• Sketch 25%

• Sketch is labeled 25%

• Explanation of strategies 25%

• Conclusions and reflections based on your results 25%

TEST GRADE:

Your completed design and the results of the test.

• Project Completed = 40%

• 50% of your grade depends on how many of your four passengers survive until the end of the track on each of your 3 runs.

• Each surviving passenger is worth 5%. That gives you a total of 12 passengers in all 3 runs (4x3) at 5% each, for a possible 60%

NOTES:

CATEGORIES: Cars, Crashes, Tracks

MISSION: Get it Running

BRIEF: You and your team have been selected to make a car that runs on rubber band power!

MISSION RULES:

1. You will design a vehicle that uses the stored energy of rubber bands to move.

2. Your device must be no more than 8 inches in any dimension.

3. You will work with one or two partners. Teams may not be of more than 3 people.

4. You must only use paper, glue, tape, rubber bands, paperclips, pencils, or other approved office supplies for your project.

5. The car may not be slingshot fired or flicked for movement, or it will result in disqualification. The car must have some sort of moving parts that rely on the use of rubber bands for movement.

QUIZ GRADE:

Create a blueprint designs for your ideas

- Sketch 25%

- Sketch is labeled 25%

- Explanation of strategies 25%

- Conclusions and reflections based on your results 25%

TEST GRADE:

Your completed design and the results of the test.

- Project Completed = 50%

- 50% of your grade depends on how far your project rolls compared to the other groups' projects. The projects that do best will get more points.

- NOTE: There is a -5% penalty for every 1/2 inch your project is out of the size specifications.

NOTES:

CATEGORIES: Cars, Distance, Rubber Bands

MISSION: Marshmallow Snowman Pileup

BRIEF: You and your team have been selected to make an MPD (marshmallow protection device). This device must be a rolling vehicle that can protect a marshmallow family as it goes down an obstacle course or simple track your teacher has developed. Track conditions MAY vary.

MISSION RULES:

1. You will design a rolling vehicle of dimensions your teacher requires that will fit on the track.

2. The vehicle must not have a completely closed top. At least 50% of the top half of the car must be open.

3. Your assembled marshmallow snowman family of 4 must be visible at all times and may not be glued or taped down to the car.

4. You may use any materials you want, provided you can scrounge them up, buy them, or find them.

5. You may test at home. In fact, you're encouraged to test at home! Assembly and design may also take place at school, but time is limited.

6. Teams may be of no more than 3 people.

TEACHER'S NOTES: This one works well with holiday snowman-shaped marshmallows. Use toothpicks to add arms and legs and markers to draw on faces. Similarly, you can run a toothpick through 2-3 mini marshmallows to make your family.

Your track can be built with simple boards, lengths of taped styrofoam, or plastic rain gutters.

QUIZ GRADE:

Create a blueprint designs for your ideas

- Labeled Sketch 25%

- Materials List 25%

- Explanation of strategies 25%

- Conclusions and reflections based on your results 25%

TEST GRADE:

Your completed design and the results of the test.

- Completed Design = 40% (penalties assessed if it does not follow the rules)

Results of Trials = up to 60%

- Each surviving family member is worth 5%.

- -5% penalty per family member that fall out of the car at any point in the track run.

- 3 trials x 4 family members = 12 lives to save, and 12 x 5% = a possible of 60%

NOTES:

CATEGORIES: Cars, Crashes, Marshmallows, Tracks

MISSION: Monster Truck Rally

BRIEF: You and your team have been selected to make the strongest monster trucks possible to crash and smash your opposition.

MISSION RULES:

1. You will design a free-rolling monster truck. It may have additional propulsion from rubber band, balloons, or other homemade type drive engines.

2. The car may be no more than 6 inches tall or wide and 9 inches long.

3. Your design must have a spot for an army man or other small driver character. Your driver may not be completely encaged or taped/glued down. There must be an adequate chance of being thrown from the car.

4. You will work with one or two partners. Teams may not be of more than 3 people.

5. You may use any approved materials to build your project. Your teacher will determine your maximum amount of materials.

TEACHER'S NOTES: A dual track design is required. Two wooden boards with shallow ledges on the sides are suggested. They should be arranged in a slight V-formation, so the cars roll toward each other and impact.

Determine your own rules and scoring method and/or a bracket system to determine final winner.

QUIZ GRADE:

A blueprint design of your idea

- Sketch 25%

- Sketch is labeled 25%

- Explanation of strategy 25%

- Conclusions and reflections based on your results 25%

TEST GRADE:

Your completed design and the results of the test.

- Project Completed = 50%

- 50% of your grade depends on how much well your project survives compared to others.

  - VICTORY CONDITIONS: a car drives over another, flips a card over, causes the other car to lose its driver, or disables a car due to damage.

  - ELIMINATION: defeated in the ways listed above, or if your car falls off of the track, does not make it down the track, flips over on its own, or is unable to continue due to self-inflicted damage.

CATEGORIES: Cars, Crashes, Survival

MISSION: Off to the Races

BRIEF:

You and your team have been selected to design a race car that will race down a teacher-designed track as fast as possible.

MISSION RULES:

1. You will design a race car that will be made from classroom supplies..

2. Your finished device must be no longer than 8 inches, no wider than 4 inches, and no taller than 4 inches.

3. You will work alone or with one or two partners. Teams may be of no more than 3 people.

4. You may use note cards, plastic straws, paper, tape, paperclips, or other approved materials in your design.

5. Time will be measured with a stopwatch as your car races down the track. You will be given 3 attempts to get the fastest time possible.

QUIZ GRADE:

Create a blueprint design for your ideas

- Sketch 25%

- Sketch is labeled 25%

- Explanation of strategies 25%

- Conclusions and reflections based on your results 25%

TEST GRADE:

Your completed design and the results of the test.

- Project Completed = 50%

- 50% of your grade depends on how fast your project runs compared to the other group's projects. The projects that do best will get more points.

- NOTE: There is a -5% penalty for every 1/2 inch your project is out of the size specifications.

NOTES:

CATEGORIES: Cars, Speed, Tracks

MISSION: Wooden Cars

BRIEF: You and your team have been selected to make a rubber-band driven wooden car.

MISSION RULES:

1. You will design a rubber-band driven wooden car that uses using spools or other gears to move.

2. The car must use rubber bands as its primary form of propulsion.

3. You will work with two or three partners. Teams may be of no more than 4 people.

4. Suggested wooden materials are: wooden thread spools (plastic may be substituted if wooden ones are not available), toothpicks, popsicle sticks, and dowel rods.

5. Up to 3 tests may be made. The car that rolls the farthest using its set number of rubber bands wins!

TEACHER'S NOTES: Wax paper is suggested as a non-sticky surface for glued pieces to dry upon.

QUIZ GRADE:

A blueprint design of your idea

- Sketch 25%

- Sketch is labeled 25%

- Explanation of strategy 25%

- Conclusions and reflections based on your results 25%

TEST GRADE:

Your completed design and the results of the test.

- Project Completed = 50%

- 50% of your grade depends on how far your car moves.

- NOTE: The car that moves the farthest gets an automatic 100%

NOTES:

CATEGORIES: Cars, Glue, Popsicle Sticks, Spools, Toothpicks, Wood

150 STEM Labs

SCIENCE * TECHNOLOGY * ENGINEERING * MATH

Eggs Projects

MISSION: Egg Survivor I - High Falls

BRIEF: You and your team have been selected to make sure a very precious cargo (a raw egg) survives a fall of at least 10 feet. Surviving eggs will then compete for increasing heights to determine which team has designed the best EPD (egg protection device).

MISSION RULES:

1. You will design a vehicle of dimensions no larger than 12x12x12. That's 12 inches long, 12 inches wide, and 12 inches deep.

2. The vehicle must have an access hole to allow us to put in an egg and then take it out to examine its condition after the fall.

3. You may use any materials you want, provided you can scrounge them up, buy them, or find them.

4. You may test at home. In fact, you're encouraged to test at home! Assembly and design may also take place at school, but time is limited.

5. Teams may be of no more than 3 people.

QUIZ GRADE:

Create a blueprint designs for your ideas

- Labeled Sketch 25%

- Materials List 25%

- Explanation of strategies 25%

- Conclusions and reflections based on your results 25%

TEST GRADE:

Your completed design and the results of the test.

- Completed Design 50% (penalties assessed if it does not follow the rules)

Results of Crash Test (up to 50%)

- Breaks on 1st attempt = 0%

- Survive 5 foot drop = 25%

- Survive 10 foot drop = 35%

- Survive Higher Fall = 50%

NOTES:

CATEGORIES: Crashes, Eggs, Height, Scavengers

MISSION: Egg Survivor II - Smallest Winner

BRIEF: You and your team have been selected to make a new EPD (egg protection device). This one must be as small as possible and must still survive a 10 foot fall!

MISSION RULES:

1. You will design a vehicle of dimensions as small as possible that can survive a 10 foot drop.

2. The vehicle must have an access hole to allow us to put in an egg and then take it out to examine its condition after the fall.

3. You may use any materials you want, provided you can scrounge them up, buy them, or find them.

4. You may test at home. In fact, you're encouraged to test at home! Assembly and design may also take place at school, but time is limited.

5. Teams may be of no more than 3 people.

QUIZ GRADE:

Create a blueprint designs for your ideas

- Labeled Sketch 25%

- Materials List 25%

- Explanation of strategies 25%

- Conclusions and reflections based on your results 25%

TEST GRADE:

Your completed design and the results of the test.

- Completed Design = 50% (penalties assessed if it does not follow the rules)

- Survives fall of 10 foot = 25% (otherwise 0%)

Size of Surviving Design (up to 25%):

- One of the smallest designs 15-25%

- One of the larger designs 0-15%

- Smallest surviving design = automatic 100%

NOTES:

CATEGORIES: Crashes, Eggs, Height, Scavengers

MISSION: Egg Survivor III - Rolling Eggs

BRIEF: You and your team have been selected to make a new EPD (egg protection device). This one must be a rolling vehicle that can protect an egg as it goes down an obstacle course or simple track your teacher has developed. Track conditions MAY vary.

MISSION RULES:

1. You will design a rolling vehicle of dimensions your teacher requires that will fit on the track.

2. The vehicle must have an access hole to allow us to put in an egg and then take it out to examine its condition after running the track.

3. You may use any materials you want, provided you can scrounge them up, buy them, or find them.

4. You may test at home. In fact, you're encouraged to test at home! Assembly and design may also take place at school, but time is limited.

5. Teams may be of no more than 3 people.

QUIZ GRADE:

Create a blueprint designs for your ideas

- Labeled Sketch 25%

- Materials List 25%

- Explanation of strategies 25%

- Conclusions and reflections based on your results 25%

TEST GRADE:

Your completed design and the results of the test.

- Completed Design = 50% (penalties assessed if it does not follow the rules)

Results of Trials = up to 50%

- Survives all three runs +50%

- Survives two runs +35%

- Survives one run +20%

- Fails first run 0%

NOTES:

CATEGORIES: Cars, Crashes, Eggs, Scavengers, Tracks

MISSION: Egg Survivor IV - Log Flume

BRIEF: You and your team have been selected to make a new EPD (egg protection device). This one must be a sailboat that can hold and protect a raw egg as it goes down a water-filled track your teacher has developed. Track conditions MAY vary.

MISSION RULES:

1. You will design a sailboat of dimensions your teacher requires that will fit on the water-filled track. A fan will push the sailboat to the end of the track. It must arrive safely.

2. The vehicle must have a place to put the egg where it is visible the entire time.

3. You may use any materials you want, provided you can scrounge them up, buy them, or find them.

4. You may test at home. In fact, you're encouraged to test at home! Assembly and design may also take place at school, but time is limited.

5. Teams may be of no more than 3 people.

QUIZ GRADE:

Create a blueprint designs for your ideas

- Labeled Sketch 25%

- Materials List 25%

- Explanation of strategies 25%

- Conclusions and reflections based on your results 25%

TEST GRADE:

Your completed design and the results of the test.

- Completed Design = 50% (penalties assessed if it does not follow the rules)

Results of Trials = up to 50%

- Survives all three runs +50%

- Survives two runs +35%

- Survives one run +20%

- Fails first run 0%

NOTES:

CATEGORIES: Boats, Crashes, Eggs, Scavengers, Tracks, Wind, Water

MISSION: Egg Survivor V - Mars Lander

BRIEF: You and your team have been selected to make a new EPD (egg protection device). This one must be a small container that protects an egg surrounded ONLY by in- flated balloons. The device must protect the egg from falls of varying heights.

MISSION RULES:

1. You will design a device of dimensions no greater than 12 inches in any direction.

2. The vehicle must have a place to put the egg where it can be checked for damage between tests.

3. You may use only a small box of card stock or other approved materials and balloons. Also acceptable for use are modest amounts of tape or glue to attach the balloons to your EPD.

4. You may test at home. In fact, you're encouraged to test at home! Assembly and design may also take place at school, but time is limited.

5. Teams may be of no more than 3 people.

QUIZ GRADE:

Create a blueprint designs for your ideas

- Labeled Sketch 25%

- Materials List 25%

- Explanation of strategies 25%

- Conclusions and reflections based on your results 25%

TEST GRADE:

Your completed design and the results of the test.

- Completed Design = 50% (penalties assessed if it does not follow the rules)

Results of Trials = up to 50%

- Breaks on 1st attempt = 0%

- Survive 5 foot drop = 25%

- Survive 10 foot drop = 35%

- Survive Higher Fall = 50%

NOTES:

CATEGORIES: Balloons, Crashes, Eggs

MISSION: Egg Survivor VI - Flak Jackets

BRIEF: You and your team have been selected to make a new EPD (egg protection device) to protect an egg in a fall. This one must only cover the egg like a jacket, and may not be more than 1 inch thick.

MISSION RULES:

1. You will design a vehicle of dimensions as small as possible that can survive as high of a drop as possible.

2. All materials must be in direct contact with the eggshell. They may not be more than 1" thick, either.

3. No matter how the egg is coated or covered, there must be a way to open it up or look to see if the egg is still intact.

4. You may use any materials you want, provided you can scrounge them up, buy them, or find them.

5. You may test at home. In fact, you're encouraged to test at home! Assembly and design may also take place at school, but time is limited.

6. Teams may be of no more than 3 people.

QUIZ GRADE:

Create a blueprint designs for your ideas

- Labeled Sketch 25%

- Materials List 25%

- Explanation of strategies 25%

- Conclusions and reflections based on your results 25%

TEST GRADE:

Your completed design and the results of the test.

- Completed Design = 50% (penalties assessed if it does not follow the rules)

- 50% of your score depends on how high of a drop it survives.

- Points will be given in 10% increments for each drop it survives: 10, 20, 30, 40, and finally 50%.

- The best project automatically gets 100%

NOTES:

CATEGORIES: Crashes, Eggs, Height, Scavengers, Survival

150 STEM Labs

SCIENCE * TECHNOLOGY * ENGINEERING * MATH

Flight Projects

MISSION: 10 Seconds and Counting...

BRIEF: You and your team have been selected to make rocket ship that can take off vertically and go as high as possible.

MISSION RULES:

1. You will design a rocket ship that is balloon-powered and attached to a string. The rocket ship must slide up along the string and go as high as possible,

2. Your rocket ship must be built from a single balloon, a straw, tape or glue, and notecards.

3. You will work with a single partner. Teams may not be of more than 2 people.

4. The straw will be used to slide along the line that is strung vertically from floor to ceiling. Both ends will be secured during tests. Inflated balloons will be attached to the rocket ship. Letting air from the balloons should propel the project as high as possible.

- TEACHERS OPTION: Have a thin rod for the plastic straw to start on, allowing the device to fly freely after clearing the launch pad. Heights must still be measured, perhaps visually against marks on an exterior wall.

QUIZ GRADE:

Research and design on jets, rockets, and propulsion.

- A paragraph on jets, engines, and propulsion 25%

- A concept idea for your rocket sketched and explained 50%

- Conclusions and reflections based on your results 25%

TEST GRADE:

Your completed design and the results of the test.

- Project Completed = 50%

- 50% of your grade depends on how high your project travels.

- NOTE: The best project gets an automatic 100%.

NOTES:

CATEGORIES: Balloons, Fliers, Height

MISSION: A Breath of Fresh Air

BRIEF: You and your team have been selected to make a rocket that can fly as high as possible on a single breath of air.

MISSION RULES:

1. You will design a rocket of any dimensions you wish.

2. You will work with one to two partners. Teams may not be of more than 3 people.

3. You may use whatever allowed materials you can find at school or home in your project.

4. Your device must be have a plastic straw tube inside of it. A second, smaller straw will be inserted into the first, and a breath of air will be blown into it to give it lift.

TEACHER'S NOTES: It is suggested that you use two different sizes of straws. The larger one should be in the rocket. The smaller one should be put inside the first, and the students will blow into the end of it to give the device lift.

Additionally, it might help if students were to all sit or lie down, so that each test was launched from a specific height. Graduated marks along a wall or measuring tapes along the wall can also help gauge heights.

QUIZ GRADE:

Create a blueprint design for your ideas

- Sketch 25%

- Sketch is labeled 25%

- Explanation of strategies 25%

- Conclusions and reflections based on your results 25%

TEST GRADE:

Your completed design and the results of the test.

- Project Completed = 50%

- 50% of your grade depends on how high your project flies compared to others.

- Top scores get +50%, and those following get +40%, +30%, or +20%.

NOTES:

CATEGORIES: Air, Fliers, Height, Plastic Straws, Scavengers

MISSION: Ben Franklin's Kites

BRIEF: You and your team have been selected to make a kite using tin foil!

MISSION RULES:

1. You will design a kite.

2. You will work with one or two partners. Teams may not be of more than 3 people.

3. The kite must be built using foil as your primary building material. It is up to your teacher how much and what kind of other materials are allowed.

4. Teams are encouraged to make non-traditional designs. Diamond-shaped kites will not get full points! Kite research is strongly recommended.

TEACHER'S NOTES: Obviously, this should not be done in poor weather. :)

QUIZ GRADE:

Create a blueprint design for your ideas

- Sketch 25%

- Sketch is labeled 25%

- Explanation of strategies 25%

- Conclusions and reflections based on your results 25%

TEST GRADE:

Your completed design and the results of the test.

- Project Completed = 50% (25% for diamond design)

- Project Gets airborne for 10+ seconds = 25%

- The remaining 25% of your points come from how long your project remains airborne compared to other teams' kites.

NOTES:

CATEGORIES: Fliers, Foil, Height, String, Wind

MISSION: Come Fly With Me

BRIEF: You and your team have been selected to design and build a kite from plastic straws and other household materials.

MISSION RULES:

1. You will design a kite.

2. You will work with one or two partners. Teams may not be of more than 3 people.

3. The kite must be built of only plastic straws, tissue paper, tape, glue, plastic wrap, and string. Your teacher will determine the amounts of each item you may use.

4. Teams are encouraged to make non-traditional designs. Diamond-shaped kites will not get full points! Kite research is strongly recommended.

QUIZ GRADE:

Create a blueprint design for your ideas

- Sketch 25%

- Sketch is labeled 25%

- Explanation of strategies 25%

- Conclusions and reflections based on your results 25%

TEST GRADE:

Your completed design and the results of the test.

- Project Completed = 50% (25% for diamond design)

- Project Gets airborne for 10+ seconds = 25%

- The remaining 25% of your points come from how long your project remains airborne compared to other teams' kites.

NOTES:

CATEGORIES: Fliers, Plastic Straws, Time

MISSION: Frisbees

BRIEF: You and your team have been selected to make a flying disk that can fly as far as possible.

MISSION RULES:

1. You will design a flying disk or ring from common classroom materials.

2. Your building materials should consist of paper, card stock, glue, rubber bands, popsicle sticks, plastic straws, and other readily-available materials that your teacher approves.

3. Your device should have a diameter of less than 18 inches.

4. You will work with alone or with a single partner. Teams may be of no more than 2 people.

5. Up to three tests will be made. Your teacher will determine how you are scored: longest shot, average, or total.

QUIZ GRADE:

A blueprint design of your idea

- Sketch 25%

- Sketch is labeled 25%

- Explanation of strategy 25%

- Conclusions and reflections based on your results 25%

TEST GRADE:

Your completed design and the results of the test.

- Project Completed = 50%

- 50% of your grade depends on how far your disk travels.

- NOTE: The best project gets an automatic 100%.

NOTES:

CATEGORIES: Distance, Fliers

MISSION: Hang Ten

BRIEF: You and your team have been selected to make a paper device that can stay in the air for the longest time possible.

MISSION RULES:

1. You will design a paper device.

2. Your finished device must be no longer in any dimension than 24 inches.

3. You will work alone or with a single partner. Teams may be of no more than 2 people.

4. You may use any thickness, color, or size paper you want, so long as the finished product is less than 24 inches in all dimensions. Additional tape, paperclips, or other approved materials may be added to your design.

5. Hang time will be measured with a stopwatch after the device is thrown. The longest time wins. Three attempts will be given.

QUIZ GRADE:

Create a blueprint designs for your ideas

- Sketch 25%

- Sketch is labeled 25%

- Explanation of strategies 25%

- Conclusions and reflections based on your results 25%

TEST GRADE:

Your completed design and the results of the test.

- Project Completed = 50%

- 50% of your grade depends on how long your project stays in the air compared to the other group's projects. The projects that do best will get more points.

- NOTE: There is a -5% penalty for every 1/2 inch your project is out of the size specifications.

NOTES:

CATEGORIES: Airplanes, Fliers, Paper, Time

MISSION: Kitetastrophe!

BRIEF: You and your team have been selected to make a variety of kites using different materials.

MISSION RULES:

1. You will design 3 kites.

2. You will work with one or two partners. Teams may not be of more than 3 people.

3. Kites should be designed using different materials like: foil, wax paper, plastic wrap, tissue paper, or cooking parchment. The frames can be assembled with bamboo skewers, plastic straws, or other lightweight materials readily available.

4. All 3 kites must take advantage of different building materials.

5. There are no size or shape restrictions, but they MUST fly!

QUIZ GRADE:

Create blueprint designs for your ideas

- Sketches 25%

- Sketches are labeled 25%

- Explanation of strategies 25%

- Conclusions and reflections based on your results 25%

TEST GRADE:

Your grade relies completely on your test results!

- Each kite is worth 33% of your grade. Each one that flies gets you that many points! Each one that does not fly costs you those 33%.

- If you get all 3 airborne for at least 10 seconds (without running to keep it aloft), you get the bonus 1% to make 100% :)

NOTES:

CATEGORIES: Fliers, Foil, Height, Paper, Plastic Wrap, String, Wax Paper, Wind

MISSION: The Landing Strip

BRIEF: You and your team have been selected to make two paper airplanes that can land as accurately on targets as possible.

MISSION RULES:

1. You will design two paper devices. One will be for long distance accuracy, while the other is for short distance accuracy.

2. Your finished devices must be no longer in any dimension than 12 inches.

3. You will work with 2 partners. Teams may be of no more than 3 people and no less than 2.

4. One teammate is responsible for the long distance test vehicle (approximately 20-30 feet), the other is responsible for the short distance test vehicle (approximately 10-15 feet). If there is a 3rd member of the team, they are the thrower. Otherwise, each builder is responsible for throwing their own plane.

5. You may use any thickness, color, or size paper you want, so long as the finished product is less than 12 inches in all dimensions. Additional tape, paperclips, or other approved materials may be added to your design.

6. There will be a total of 5 attempts to land the airplanes on/in their targets at each distance. Each successful landing is worth a point. Close targets might be landing on a Frisbee. Long distance targets might be landing inside a hula hoop (or flying through it). This is the teacher's discretion.

QUIZ GRADE:

Create two blueprint designs for your ideas

- 2 Sketches 25%

- Sketches are labeled 25%

- Explanation of strategies 25%

- Conclusions and reflections based on your results 25%

TEST GRADE:

Your completed design and the results of the test.

- Projects Completed = 50% (25% each)

- 50% of your grade depends on how many points your team can accumulate compared to the other groups. The teams that gather the most points will get more points, with 1st place automatically getting 100%.

NOTES:

CATEGORIES: Accuracy, Airplanes, Fliers, Paper

MISSION: Lay it All on the Line

BRIEF: You and your team have been selected to make a rocket ship that races down a hanging line.

MISSION RULES:

1. You will design a rocket ship that is balloon-powered and attached to a string. The rocket ship must slide along the string and go as far as possible,

2. Your rocket ship must be built from a single balloon, a straw, tape or glue, and notecards.

3. You will work with a single partner. Teams may not be of more than 2 people.

4. The straw will be used to slide along the line that is strung across the room. Both ends will be secured during tests. Inflated balloons will be attached to the rocket ship. Letting air from the balloons should propel the project as far as possible.

QUIZ GRADE:

Research and design on jets, rockets, and propulsion.

- A paragraph on jets, engines, and propulsion 25%

- A concept idea for your rocket sketched and explained 50%

- Conclusions and reflections based on your results 25%

TEST GRADE:

Your completed design and the results of the test.

- Projects Completed = 50% (25% each)

- 50% of your grade depends on how far your project travels. There may be up to 3 trials for you to get the best results possible.

- NOTE: The best project gets an automatic 100%.

NOTES:

CATEGORIES: Balloons, Distance, Fliers, String

MISSION: Look Out Below!

BRIEF: You and your team have been selected to make a parachute that slows a weight down as much as possible, taking the longest time to reach the floor from the drop point.

MISSION RULES:

1. You will design a device to slow the descent of a weight your teacher has chosen (a graduated weight, a ping pong ball, army man, etc...) as much as possible

2. Your device may be built from any approved materials you can find at home or in the class, including: straws, tissue paper, tape or glue, foil, strings notecards...

3. You will work with one or two partners. Teams may be of no more than 3 people.

4. Your device will be tested from a set height of 6-10 feet, or higher if possible.

5. Your device must have some sort of hook or basket for the weight to be attached to your project. The requirements for this depend on what weighted object your teacher has selected.

6. There will be up to 3 tests with small adjustments allowed between testings.

TEACHER'S NOTES: Use of a stopwatch and a ladder may make this project better. Awards can be given for slowest time (which is best) and best average.

QUIZ GRADE:

Create a blueprint designs for your ideas

- Sketch 25%

- Sketch is labeled 25%

- Explanation of strategies 25%

- Conclusions and reflections based on your results 25%

TEST GRADE:

Your completed design and the results of the test.

- Project Completed = 50%

- 50% of your grade depends on how long your project stays in the air compared to the other group's projects. The projects that do best will get more points.

- NOTE: The project that takes the longest to land gets an automatic 100%

NOTES:

CATEGORIES: Fliers, Scavengers, Time

MISSION: Sticky Planes

BRIEF: You and your team have been selected to make as paper airplanes that stick to a wall.

MISSION RULES:

1. You will design a paper airplane that flies across a specified distance and sticks to a target on a wall.

2. The wall may be treated with velcro or tape for stickiness. The design will have to be made in such a way as to hit the wall and stick.

3. No dimension of your airplane should be over 18 inches.

4. You will work with one or two partners. Teams may not be of more than 3 people.

5. Your teacher will determine the amount of and type of materials you receive.

6. There will be 3 attempts, possibly at different distances.

TEACHER'S NOTES: Students might be given a piece of the hook side of velcro strip, and there may be a target made of fuzzy felt on the wall. Additional options include flypaper and magnetic strips.

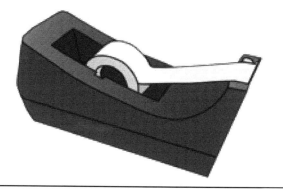

QUIZ GRADE:

A blueprint design of your idea

- Sketch 25%

- Sketch is labeled 25%

- Explanation of strategy 25%

- Conclusions and reflections based on your results 25%

TEST GRADE:

Your completed design and the results of the test.

- Project Completed = 40%

- Project sticks 1 time = +20%

- Project sticks 2 times = +40%

- Project sticks 3 times = +60%

NOTES:

CATEGORIES: Accuracy, Fliers, Paper, Tape, Velcro

MISSION: To the Moon and Back

BRIEF: You and your team have been selected to make a paper device that can fly the greatest distance in a straight line from a launch point.

MISSION RULES:

1. You will design a paper device.

2. Your finished device must be no longer in any dimension than 24 inches.

3. You will work alone or with a single partner. Teams ma y be of no more than 2 people.

4. You may use any thickness, color, or size paper you want, so long as the finished product is less than 24 inches in all dimensions. Additional tape, paperclips, or other approved materials may be added to your design.

5. Distance will be measured with a tape measure or other measuring tool after the device is thrown. The longest distance wins. Three attempts will be given.

QUIZ GRADE:

A blueprint design of your idea

- Sketch 25%

- Sketch is labeled 25%

- Explanation of strategy 25%

- Conclusions and reflections based on your results 25%

TEST GRADE:

Your completed design and the results of the test.

- Project Completed = 50%

- 50% of your grade depends on how far your project flies compared to the other group's projects. The projects that do best will get more points.

- NOTE: There is a -5% penalty for every 1/2 inch your project is out of the size specifications.

NOTES:

CATEGORIES: Airplanes, Distance, Fliers, Paper

MISSION: We Can Fly Anything!

BRIEF:
You and your team have been selected to make a flying device that can fly the greatest distance in a straight line from a launch point. The catch is, you don't get to choose the material you use!

MISSION RULES:

1. You will design a paper device after being given a randomly-selected piece of building material, which may include but is not limited to: parchment paper, construction paper, tissue paper, tin foil, wax paper, card stock, and writing paper.

2. Your teacher will help teams randomly select different materials for different rounds. Each team will get a different material for each round, never the same materials twice. At the end of all rounds, every team will have used each material.

3. You only get one piece of building material per round, so consider your strategies before making the device, and you may not simply wad up the material into a ball and throw it.

4. You will work alone or with a single partner. Teams may be of no more than 2 people.

5. Distance will be measured with a tape measure or other measuring tool after the device is thrown. The longest distance wins. Three attempts will be given.

QUIZ GRADE:

Plans for each of your projects

- Each project for each material must have a strategy . Your strategies should reflect how you change your designs for the different building materials. 75%

- Conclusions and reflections based on your results 25%

TEST GRADE:

Your completed design and the results of the test.

- Projects are Completed = 75%

- 25% of your grade depends on how far your projects fly compared to the other group's projects. This scoring is done as a total distance of ALL rounds once completed. The projects that do best will get more points.

- NOTE: Because there are multiple rounds, one per each building material, there are NO redo's. It is one attempt per device per round.

NOTES:

CATEGORIES: Airplanes, Distance, Fliers, Paper

MISSION: Whirly Birds

BRIEF: You and your team have been selected to make a homemade helicopter that can take off from a stationary position.

MISSION RULES:

1. You will design a helicopter that can take off from the ground.

2. Your helicopter must be built from rubber bands, straws, toothpicks, notecards, popsicle sticks, glue, and other classroom items.

3. You will work with a single partner. Teams may not be of more than 2 people.

4. Using the rubber bands to power the propellers, the frame and the body of the helicopter should be able to lift off at least temporarily.

QUIZ GRADE:

Research and/or design on helicopters.

- A paragraph on helicopters, especially early ones like DaVinci's designs 25%

- A concept idea for your helicopter sketched and explained 50%

- Conclusions and reflections based on your results 25%

TEST GRADE:

Your completed design and the results of the test.

- Project Completed = 50%

The success of your project determines the other 50% of your grade.

- +10% = didn't move at all.

- +25% = almost took off

- +50% = completely lifted off the ground.

NOTES:

CATEGORIES: Fliers, Height, Rubber Bands

150 STEM Labs

SCIENCE * TECHNOLOGY * ENGINEERING * MATH

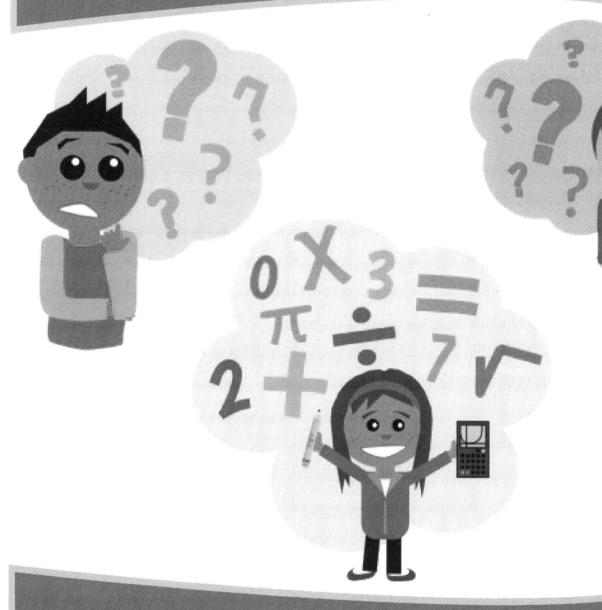

Machines Projects

MISSION: Block and Tackle

BRIEF: You and your team have been selected to make a wooden block and tackle.

MISSION RULES:

1. You will design a block and tackle using just dowel rods, thread spools, popsicle sticks, and glue.

2. The spools must be able to spin and turn with the movement of the string or line.

3. You will work with one or two partners. Teams may be of no more than 3 people.

4. Your device must have paperclips to anchor the top and to hook weights and strings to.

5. Weight will be gradually added to the system, until it breaks or is near breaking. Holding more weight means a better score.

TEACHER'S NOTES: Wax paper is suggested as a non-sticky surface for glued pieces to dry upon.

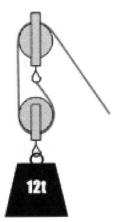

QUIZ GRADE:

A blueprint design of your idea

- Sketch 25%

- Sketch is labeled 25%

- Explanation of strategy 25%

- Conclusions and reflections based on your results 25%

TEST GRADE:

Your completed design and the results of the test.

- Project Completed = 50%

- 50% of your grade depends on how much weight your device holds before breaking.

- NOTE: The car that moves the farthest gets an automatic 100%

NOTES:

CATEGORIES: Glue, Paperclips, Popsicle Sticks, Spools, String, Weight, Wood

MISSION: Construction Chaos I - Excavator

BRIEF: You and your team have been selected to make an excavator machine to transport objects.

MISSION RULES:

1. You will design an excavator truck (backhoe, bucket truck...) to pick up a load and move it to a designated location.

2. Your machine may be any dimensions, but it must roll and it must be able to pick up the loads specified by your teacher.

3. You may not use your hands to directly touch the payload. You may use your hands to roll your machine and manipulate levers, pulleys, and other parts of your machine to complete your tasks, though.

4. There will be 3 loads. Each of them may vary in dimensions and weights. You must design a versatile machine that can scoop/pick up the loads.

5. You will work with one or two partners. Teams may not be of more than 3 people.

6. You may use any approved materials you can find at school or at home, including paper clips, pipe cleaners, plastic straws, notecards, tape...

TEACHER'S NOTES: Generally, moving a small load, like a AA battery, a coin, or a paperclip from one end of the table to the other without being touched by hands is the goal. Test 3 different loads or the same load 3 times.

QUIZ GRADE:

Research and design on construction equipment designs.

- A paragraph on the parts and designs of construction equipment 25%

- A concept idea for your machine sketched and explained 50%

- Conclusions and reflections based on your results 25%

TEST GRADE:

Your completed design and the results of the test.

- Project Completed = 25%

- 75% of your grade depends on if your project actually works. Each successful delivery of the load is worth 25%

NOTES:

CATEGORIES: Accuracy, Gears, Levers, Pulleys, Scavengers

MISSION: Construction Chaos II - Crane

BRIEF:

You and your team have been selected to make an crane to lift and move objects.

MISSION RULES:

1. You will design a stationary crane to pick up a load and move it to a designated location.

2. Your machine may be any dimensions, but it must must be able to pick up the loads specified by your teacher.

3. You may not use your hands to directly touch the payload. You may use your hands to manipulate levers, pulleys, and other parts of your machine to complete your tasks, though.

4. There will be 3 loads. Each of them may vary in dimensions and weights. You must design a versatile machine that can scoop/hook/pick up the loads.

5. You will work with one or two partners. Teams may not be of more than 3 people.

6. You may use any approved materials you can find at school or at home, including paper clips, pipe cleaners, plastic straws, notecards, tape...

TEACHER'S NOTES: All 3 objects should have a hook or loop on them that can be caught. Varying weights will test the strength and durability of the machines.

QUIZ GRADE:

Research and design on construction equipment designs.

- A paragraph on the parts and designs of construction equipment 25%

- A concept idea for your machine sketched and explained 50%

- Conclusions and reflections based on your results 25%

TEST GRADE:

Your completed design and the results of the test.

- Project Completed = 25%

- 75% of your grade depends on if your project actually works. Each successful delivery of the load is worth 25%

NOTES:

CATEGORIES: Accuracy, Gears, Levers, Pulleys, Scavengers

MISSION: Construction Chaos III - Forklift

BRIEF: You and your team have been selected to make an forklift machine to transport objects.

MISSION RULES:

1. You will design forklift to pick up a load and move it to a designated location.

2. Your machine may be any dimensions, but it must roll and it must be able to pick up the loads specified by your teacher.

3. You may not use your hands to directly touch the payload. You may use your hands to roll your machine and manipulate levers, pulleys, and other parts of your machine to complete your tasks, though.

4. There will be 3 loads. Each of them may vary in dimensions and weights. You must design a versatile machine that can scoop/pick up the loads.

5. Each load MUST be picked up, not just skidded and pushed across the table surface.

6. You will work with one or two partners. Teams may not be of more than 3 people.

7. You may use any approved materials you can find at school or at home, including paper clips, pipe cleaners, plastic straws, notecards, tape...

TEACHER'S NOTES: Small, flat objects work best for loads. Little memo pads, a baseball card, etc...

QUIZ GRADE:

Research and design on construction equipment designs.

- A paragraph on the parts and designs of construction equipment 25%

- A concept idea for your machine sketched and explained 50%

- Conclusions and reflections based on your results 25%

TEST GRADE:

Your completed design and the results of the test.

- Project Completed = 25%

- 75% of your grade depends on if your project actually works. Each successful delivery of the load is worth 25%

NOTES:

CATEGORIES: Accuracy, Gears, Levers, Pulleys, Scavengers

MISSION: Flippers

BRIEF: You and your team have been selected to make a device that can flip things over.

MISSION RULES:

1. You will design a device that can flip over as much weight as possible.

2. Your device should not be over 12 inches in any dimension.

3. You will work with one to two partners. Teams may not be of more than 3 people.

4. You may use any approved materials from home or school to make your device.

5. Your device must be free-standing and movable. It cannot be attached to any surface.

6. Your device should have some sort of skid that slides under the books or weight to be flipped.

7. Your device should have a lever or activator that activates your machine's flipping action.

TEACHER'S NOTES: It is suggested that you use paperback books for flipping. The devices will be activated and should flip the books over.

QUIZ GRADE:

Create a blueprint design for your ideas

- Sketch 25%

- Sketch is labeled 25%

- Explanation of strategies 25%

- Conclusions and reflections based on your results 25%

TEST GRADE:

Your completed design and the results of the test.

- Project Completed = 50%

- 50% of your grade depends on how much weight your project can flip over compared to others.

- Top scores get +50%, and those following get +40%, +30%, or +20%.

NOTES:

CATEGORIES: Levers, Scavengers, Weight

MISSION: Flushers

BRIEF: You and your team have been selected to make a working toilet.

MISSION RULES:

1. You will design a device that can flush dirty water into a holding tank, replacing it with fresh water.

2. Your device should not be over 18 inches in any dimension.

3. You will work with two to three partners. Teams may not be of more than 4 people.

4. You may use any approved materials from home or school to make your device.

5. Your device must be free-standing and movable. It cannot be attached to any surface.

6. Your device should have a lever or activator that activates your machine's flushing action.

7. Your device should have 3 separate holding areas: the fresh water tank, the bowl, and the dirty water tank. Water should move between then upon activating your machine.

TEACHER'S NOTES: It is suggested that you use food coloring or plastic beads to simulate the 'dirty' water and to determine if it is being cleaned as it is drained and replaced.

QUIZ GRADE:

Create a blueprint design for your ideas

- Sketch 25%

- Sketch is labeled 25%

- Explanation of strategies 25%

- Conclusions and reflections based on your results 25%

TEST GRADE:

Your completed design and the results of the test.

- Project Completed = 50%

- 50% of your grade depends on how well your project works.

  - Project replaces water +25%

  - Project cleans water +15%

  - Project can be flushed twice +10%

NOTES:

CATEGORIES: Bottles, Cups, Scavengers, Water

MISSION: Gone Fishin'

BRIEF: You and your team have been selected to make a working fishing pole.

MISSION RULES:

1. You will design a working fishing pole. It will wind up cord when a hand crank is turned.

2. The fishing pole and reel may be built from any approved materials found at home or at school. Suggested materials are: plastic straws, cardboard, rubber bands, paperclips, glue, tape, thread spools, etc...

3. You will work with one or two partners. Teams may be of no more than 3 people.

4. Your device's reel should be less than 6 inches in all dimensions. The pole may be up to 3 feet long.

5. The pole must have eyelets to thread the line through, and the line must have a paperclip hook on the end.

6. The pole will be strength-tested to see how large of a 'fish' or weight it can pull in.

TEACHER'S NOTES: Suggested 'fish' are graduated weights with eyelets on them. Toy fish of varying weights can also be used.

Reels don't have to cast, but they could. They should at least allow the line to be pulled back out.

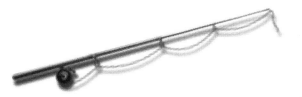

QUIZ GRADE:

A research paper on fishing reels.

- 2-3 pictures of fishing reels 25%

- A labeled concept idea based on your fishing reel pictures, including what materials you hope to use for each piece 50%

- Conclusions and reflections based on your results 25%

TEST GRADE:

Your completed design and the results of the test.

- Project Completed = 50%

- Reel works = 20%

- 30% of your grade depends on how much weight your reel and pole can handle compared to other projects.

NOTE: The best project gets 100%

NOTES:

CATEGORIES: Paperclips, Plastic Straws, Scavengers, Spools, String, Weight

MISSION: Hand Mixer

BRIEF: You and your team have been selected to make a working hand mixer.

MISSION RULES:

1. You will design a working hand mixer. It will use gears or some sort of drive to turn the mixer(s) when you turn the hand crank.

2. The mixer may be built from any approved materials found at home or at school. Suggested materials are: plastic straws, cardboard, rubber bands, paperclips, glue, tape, etc...

3. You will work with one or two partners. Teams may be of no more than 3 people.

4. Your device may be of any dimensions less than 12 inches.

TEACHER'S NOTES: Style and Function awards are a cool option.

QUIZ GRADE:

A research paper on mixers.

- 2-3 pictures of hand crank mixers 25%

- A labeled concept idea based on your mixer pictures, including what materials you hope to use for each piece 50%

- Conclusions and reflections based on your results 25%

TEST GRADE:

Your completed design and the results of the test.

- Project Completed = 50%

- 50% of your grade depends on how well your project works to stir a bowl of water.

  - Works: Well +50%, Okay 30%, Hardly at all, 10%

  - More than 1 set of mixer tines: +20%

NOTES:

CATEGORIES: Gears, Paper Clips, Scavengers, Water

MISSION: Marshmallow Mayhem

BRIEF: You and your team have been selected to make a device that can crush marshmallows as flat as possible.

MISSION RULES:

1. You will design a device that will crush a regular-sized marshmallow as flat as possible.

2. 3 seconds after being hit/crushed/ smashed, the marshmallow will be measured for height at the tallest point.

3. You may use any approved materials that you find in the classroom or at home.

4. You may not simply drop a weight on the marshmallow. There must be a device/ lever/mechanism that creates an action that crushes the marshmallow.

5. You will work with one to two partners. Teams may not be of more than 3 people.

6. 3 tests will be made. Your teacher may decide on a best of 3, an average score, or some composite score.

QUIZ GRADE:

Create a blueprint design for your ideas

• Sketch 25%

• Sketch is labeled 25%

• Explanation of strategies 25%

• Conclusions and reflections based on your results 25%

TEST GRADE:

Your completed design and the results of the test.

• Project Completed = 50%

• 50% of your grade depends on how flattened your marshmallow is compared to others.

• Top scores get +50%, and those following get +40%, +30%, or +20%.

NOTES:

CATEGORIES: Crashes, Marshmallows, Scavengers

MISSION: Merry-Go-Round!

BRIEF: You and your team have been selected to make a balloon-powered merry-go-round!

MISSION RULES:

1. You will design a Merry-Go-Round that uses balloon power to turn about a central axis.

2. Your device must have an arm or flag of sorts that protrudes out far enough to make it visually obvious, so that the number of successful rotations can be counted when your device is in motion.

3. Your device must anchor to a surface like a desk and rotate about a central pole or pin.

4. You may only blow up your balloon once per trial. You may decide how full you want the balloon to be.

5. Design your project from whatever approved materials you can find at home or in the classroom.

6. You will work with 1-2 partners. Teams may not be of more than 3 people.

7. More successful rotations means a better score.

QUIZ GRADE:

Create a blueprint design for your ideas

- Sketch 25%

- Sketch is labeled 25%

- Explanation of strategies 25%

- Conclusions and reflections based on your results 25%

TEST GRADE:

Your completed design and the results of the test.

- Project Completed = 50%

- 50% of your grade depends on how many times your project rotates compared to other projects.

- You will add up the total revolutions in 3 trials, with only minor adjustments and fixes between trials. The project with the most rotations automatically gets 100%

NOTES:

CATEGORIES: Balloons, Scavengers

MISSION: Pinball Fever

BRIEF: You and your team have been selected to make a pinball machine for a marble, which will take as long as possible for the marble to return to the launch.

MISSION RULES:

1. You will research pinball machines and get ideas for a concept for your design.

2. Your design may be of any dimensions less than 3 feet in any one direction.

3. You will work with two or three partners. Teams may not be of more than 4 people.

4. You may use any approved scavenged materials from home and school to build your device.

5. Your device will use a clothespin to launch the marble.

6. The marble will go through a series of tricks or structures to delay the return of the marble to the start point. The longer it takes to return, the better.

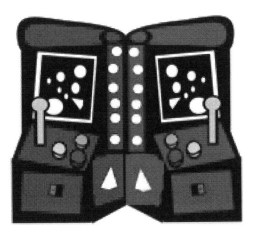

QUIZ GRADE:

A research paper on pinball machines.

- 2-4 pictures of pinball machines 25%

- A concept idea for your pinball machine 50%

- Conclusions and reflections based on your results 25%

TEST GRADE:

Your completed design and the results of the test.

- Project Completed = 50%

- 50% of your grade depends on how long the marble rolls compared to the other group's projects. The projects that do best will get more points.

- NOTE: There is a -5% penalty for each time the marble gets stuck. You get 1 free restart.

NOTES:

CATEGORIES: Clothespins, Marbles, Scavengers, Time

MISSION: Scavenger Bikes

BRIEF: You and your team have been selected to make a working miniature version of a bicycle from scavenged materials.

MISSION RULES:

1. You will design a working bike or tricycle. It must have a chain drive or belt drive to allow the pedals to move the rear wheel.

2. The bike may be built from any approved materials found at home or at school. Suggested materials are: cardboard tubes, plastic straws, cardboard, rubber bands, paperclips, glue, tape, etc...

3. You will work with one or two partners. Teams may be of no more than 3 people.

4. Your device may be of any dimensions less than 18 inches.

TEACHER'S NOTES: Style and Function awards are a cool option.

QUIZ GRADE:

A research paper on bikes.

- 2-4 pictures of bikes 25%

- A labeled concept idea based on your bike pictures, including what materials you hope to use for each piece 50%

- Conclusions and reflections based on your results 25%

TEST GRADE:

Your completed design and the results of the test.

- Project Completed = 50%

- 50% of your grade depends on whether or not the belt/chain drive works for the rear tire.

NOTE: Adding brakes or other details may help your score.

NOTES:

CATEGORIES: Bikes, Gears, Rubber Bands, Scavengers

MISSION: Splashing Around

BRIEF: You and your team have been selected to make a water wheel that turns with moving water and performs one or more actions.

MISSION RULES:

1. You will design and build a water wheel from notecards, glue, tape, toothpicks, popsicle sticks, rubber bands, straws, or other scavenged household materials.

2. Your finished water wheel must turn as result of moving water, like the water from a faucet.

3. Your finished water wheel must have gears or mechanisms that turn or make at least one evident and visible action occur, like raising and lowering a flag, pushing something, etc...

4. You will work with two or three partners. Teams may be of no more than 4 people.

TEACHERS' NOTES: This one works best over by a faucet. Devices would have to be designed to sit on a counter, which a portion of the device (the water wheel and some sort of driveshaft) extending out to reach the running water.

QUIZ GRADE:

A blueprint design of your idea

- Sketch 25%

- Sketch is labeled 25%

- Explanation of strategy 25%

- Conclusions and reflections based on your results 25%

TEST GRADE:

Your completed design and the results of the test.

- Project Completed = 50%

50% of your grade depends on what your project does as an action(s).

- One action = +20%

- Two actions = +35%

- Three actions = +50%

NOTES:

CATEGORIES: Gears, Rotation, Scavengers, Water

MISSION: Take to the Winds

BRIEF: You and your team have been selected to make a working windmill that turns gears and makes an action occur as a result of the windmill and gears turning.

MISSION RULES:

1. You will design and build a windmill from notecards, glue, tape, toothpicks, popsicle sticks, rubber bands, straws, or other scavenged household materials.

2. Your finished windmill must turn in the wind.

3. Your finished windmill must have gears inside that turn or make at least one evident and visible action occur, like raising and lowering a flag, pushing something, etc...

4. You will work with two or three partners. Teams may be of no more than 4 people.

QUIZ GRADE:

A blueprint design of your idea

- Sketch 25%

- Sketch is labeled 25%

- Explanation of strategy 25%

- Conclusions and reflections based on your results 25%

TEST GRADE:

Your completed design and the results of the test.

- Project Completed = 50%

50% of your grade depends on what your project does as an action(s).

- One action = +20%

- Two actions = +35%

- Three actions = +50%

NOTES:

CATEGORIES: Devices, Machines, Scavengers, Wind

MISSION: Wheel of Fortune

BRIEF: You and your team have been selected to make a rotating ferris wheel that is powered by rubber bands.

MISSION RULES:

1. You will design a ferris wheel that is powered by rubber bands.

2. Your ferris wheel must be at least 6 inches in diameter.

3. You will work with one or two partners. Teams may not be of more than 3 people.

4. You may use any approved materials you can find at school or at home, including paper clips, pipe cleaners, plastic straws, notecards, tape...

TEACHER'S NOTES: You can add extra difficulty, like making at least four seats or boxes for army men. Each army man must make at least one rotation/revolution through the ride for it to be considered a success.

QUIZ GRADE:

Research and design on ferris wheels and thrill rides.

- A paragraph on ferris wheels 25%

- A concept idea for your ferris wheel sketched and explained 50%

- Conclusions and reflections based on your results 25%

TEST GRADE:

Your completed design and the results of the test.

- Project Completed = 50%

- 50% of your grade depends on if your project actually works. More rotations = better grade.

- NOTE: The best project gets an automatic 100%.

NOTES:

CATEGORIES: Gears, Rotation, Rubber Bands, Scavengers

150 STEM Labs

SCIENCE * TECHNOLOGY * ENGINEERING * MATH

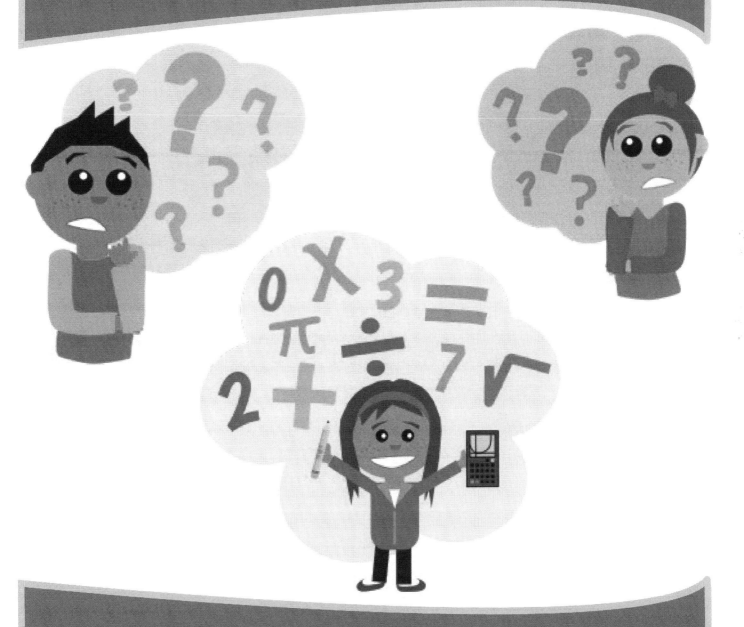

Strength Test Projects

MISSION: All Tangled Up

BRIEF: You and your team have been selected to make a the best knot possible.

MISSION RULES:

1. You will design and tie a knot that will tie to a pole and to a bucket.

2. Your knot must be something you and your partner can both demonstrate. One of you will tie the knot to the pole, and the other will tie the knot to the bucket.

3. You will work with one partner. Teams may not be of more than 2 people.

4. You must only a single string of a length set by your teacher for your project. You will be given several short practice lengths, though, so you can perfect your knots.

5. Once the knots are tied, the pole will be set across two desks or tables, and the bucket will be suspended from the string. Weight will be added to the bucket until the string snaps.

QUIZ GRADE:

- Written step-by-step knot-tying procedures 25%

- Demonstration of knot tying by both partners 25%

- Explanation of strategy 25%

- Conclusions and reflections based on your results 25%

TEST GRADE:

Your completed design and the results of the test.

- Project Completed = 50%

- 50% of your grade depends on how much weight your knot holds compared to the other group's projects. The projects that do best will get more points.

NOTES:

CATEGORIES: Chains, Materials Strength, String, Weight

MISSION: Can Crusher

BRIEF: You and your team have been selected to make the strongest device from ONLY tin foil to hold up as much weight as possible.

MISSION RULES:

1. You will design a device that holds a small container on top. The container will be filled gradually with weight (sand, paperclips, marbles, or coins), and must hold as much weight as possible.

2. Your teacher will determine the amount of tin foil you may use in your project. You may use only tin foil for your construction.

3. You will work with one to two partners. Teams may not be of more than 3 people.

4. Your device must be free-standing and movable. It cannot be attached to any surface.

QUIZ GRADE:

Create a blueprint design for your ideas

- Sketch 25%

- Sketch is labeled 25%

- Explanation of strategies 25%

- Conclusions and reflections based on your results 25%

TEST GRADE:

Your completed design and the results of the test.

- Project Completed = 50%

- 50% of your grade depends on how much weight your project can hold up compared to others.

- Top scores get +50%, and those following get +40%, +30%, or +20%.

NOTES:

CATEGORIES: Dead Lift, Foil, Materials Strength, Weight

MISSION: Daisy Chains

BRIEF: You and your team have been selected to make the strongest chain possible from strips of paper!

MISSION RULES:

1. You will design a paper chain at least 24 inches long, and your chain must be created from paper and one type of the following adhesives: tape, glue, or staples. Your teacher will decide on your maximum number of sheets of paper and the amount of adhesives you get to use.

2. Your chain must attach at both ends to test its weight allowance.

3. Your teacher will have a container to attach to one end of your chain and a pole, rope loop, or metal eyelet to attached to the other .

4. You must work with only 1 partner, or alone. Teams may not be made up of more than 2 people.

QUIZ GRADE:

Create a blueprint design for your ideas

- Sketch 25%

- Sketch is labeled 25%

- Explanation of strategies 25%

- Conclusions and reflections based on your results 25%

TEST GRADE:

Your completed design and the results of the test.

- Project Completed = 40%

50% of your grade depends on your chain's success in holding weight.

- Projects will get between 10-50% depending on the amount of weight they hold.

- NOTE: The best project gets an automatic 100%.

NOTES:

CATEGORIES: Chains, Materials Strength, Paper, Weight

MISSION: Dead Lift

BRIEF: You and your team have been selected to make the strongest possible book holder out of plastic straws and tape.

MISSION RULES:

1. You will design a device at least 6 inches tall with any other dimensions of length and width that can hold up as many textbooks as possible without collapsing.

2. Your teacher will determine the maximum number of straws you may use in your project.

3. You will work with one to two partners. Teams may not be of more than 3 people.

4. You may use only plastic straws and clear tape for your construction.

5. Your device must be free-standing and movable. It cannot be attached to any surface.

QUIZ GRADE:

Create a blueprint design for your ideas

- Sketch 25%

- Sketch is labeled 25%

- Explanation of strategies 25%

- Conclusions and reflections based on your results 25%

TEST GRADE:

Your completed design and the results of the test.

- Project Completed = 50%

- 50% of your grade depends on how much weight your project can hold up compared to others.

- Top scores get +50%, and those following get +40%, +30%, or +20%.

NOTES:

CATEGORIES: Dead Lift, Materials Strength, Plastic Straws, Weight

MISSION: Glue-ten Free

BRIEF: You and your team have been selected to make the strongest possible structure out of glue and string or yarn.

MISSION RULES:

1. You will design a device that is at least 4 inches in all three dimensions, which can hold up as much weight as possible without collapsing.

2. Your teacher will determine the maximum amount of materials you may use in your project.

3. You will work with one to two partners. Teams may not be of more than 3 people.

4. You may use only string/yarn and glue for your construction.

5. Your device must be free-standing and movable. It cannot be attached to any surface.

TEACHER'S NOTES: It is suggested that you use wax paper to set up 'cables' of thread or yarn. Glue will help them set into a specific shape when it dries. Then they can be layered and attached to form the structure.

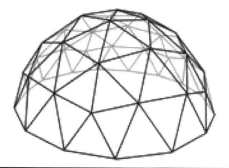

QUIZ GRADE:

Create a blueprint design for your ideas

- Sketch 25%

- Sketch is labeled 25%

- Explanation of strategies 25%

- Conclusions and reflections based on your results 25%

TEST GRADE:

Your completed design and the results of the test.

- Project Completed = 50%

- 50% of your grade depends on how much weight your project can hold up compared to others.

- Top scores get +50%, and those following get +40%, +30%, or +20%.

NOTES:

CATEGORIES: Dead Lift, Glue, Materials Strength, String, Weight, Yarn

MISSION: Huff and Puff Your House Down

BRIEF: You and your team have been selected to make a house out of plastic straws and other basic building materials that can withstand a wrecking ball's impending attack.

MISSION RULES:

1. You will design a house that is at least 6 inches on a side. It may be cubic or more rectangular, but it must have some noticeable features of a house, like a door, a roof, windows, etc...

2. You may use only plastic straws, clear tape, and note cards for your construction. Your teacher will determine the maximum number of straws and other supplies you may use in your project.

3. You will work with one to two partners. Teams may not be of more than 3 people.

4. Your device must be free-standing and movable. It cannot be attached to any surface.

5. Your device will be repeatedly struck with a tennis ball or other object on a pendulum string. It must survive as many attacks as possible.

QUIZ GRADE:

Create a blueprint designs for your ideas

- Sketch 25%

- Sketch is labeled 25%

- Explanation of strategies 25%

- Conclusions and reflections based on your results 25%

TEST GRADE:

Your completed design and the results of the test.

- Project Completed = 50%

- 50% of your grade depends on how well your project withstands to attacks: Survive 1 attack +25%, 2 attacks +35%, 3 attacks +45%, 5 attacks or more +50%

- NOTE: Your teacher will decide if your house has 'survived' or not, depending on damage sustained.

NOTES:

CATEGORIES: Crashes, Materials Strength, Notecards, Plastic Straws

MISSION: On a Strong Note...

BRIEF: You and your team have been selected to make the strongest device from ONLY note cards to hold up as much weight as possible.

MISSION RULES:

1. You will design a device that holds a small container on top. The container will be filled gradually with weight (sand, paperclips, marbles, or coins), and must hold as much weight as possible.

2. Your teacher will determine the maximum number of note cards you may use in your project. You may use only note cards for your construction, unless otherwise specified. Your teacher may or may not allow a small amount of adhesives.

3. You will work with one to two partners. Teams may not be of more than 3 people.

4. Your device must be free-standing and movable. It cannot be attached to any surface.

5. OPTIONAL: Use Post-It's instead of note cards, so the stickiness offers more design options.

QUIZ GRADE:

Create a blueprint design for your ideas

- Sketch 25%

- Sketch is labeled 25%

- Explanation of strategies 25%

- Conclusions and reflections based on your results 25%

TEST GRADE:

Your completed design and the results of the test.

- Project Completed = 50%

- 50% of your grade depends on how much weight your project can hold up compared to others.

- Top scores get +50%, and those following get +40%, +30%, or +20%.

NOTES:

CATEGORIES: Dead Lift, Materials Strength, Notecards, Weight

MISSION: Pipe Dreams

BRIEF: You and your team have been selected to make the strongest device from ONLY pipe cleaners to hold up as much weight as possible.

MISSION RULES:

1. You will design a device that holds a small container on top. The container will be filled gradually with weight (sand, paperclips, marbles, or coins), and must hold as much weight as possible.

2. Your teacher will determine the maximum number of pipe cleaners you may use in your project. You may use only pipe cleaners for your construction.

3. You will work with one to two partners. Teams may not be of more than 3 people.

4. Your device must be free-standing and movable. It cannot be attached to any surface.

QUIZ GRADE:

Create a blueprint design for your ideas

- Sketch 25%

- Sketch is labeled 25%

- Explanation of strategies 25%

- Conclusions and reflections based on your results 25%

TEST GRADE:

Your completed design and the results of the test.

- Project Completed = 50%

- 50% of your grade depends on how much weight your project can hold up compared to others.

- Top scores get +50%, and those following get +40%, +30%, or +20%.

NOTES:

CATEGORIES: Dead Lift, Materials Strength, Pipe Cleaners, Weight

MISSION: Pump You Up

BRIEF: You and your team have been selected to make the strongest possible book holder out of paper tubes and glue.

MISSION RULES:

1. You will design a device at least 6 inches tall with any other dimensions of length and width that can hold up as many textbooks as possible without collapsing.

2. Your teacher will determine the maximum number of cardboard tubes you may use in your project.

3. You will work with one to two partners. Teams may not be of more than 3 people.

4. You may use only cardboard tubes and glue for your construction.

5. Your device must be free-standing and movable. It cannot be attached to any surface.

TEACHER'S NOTES: You might want to have students start collecting cardboard tubes from paper towels and gift wrapping paper a good deal ahead of time before they start the project. Toilet paper rolls are likely not sanitary...

QUIZ GRADE:

Create a blueprint design for your ideas

• Sketch 25%

• Sketch is labeled 25%

• Explanation of strategies 25%

• Conclusions and reflections based on your results 25%

TEST GRADE:

Your completed design and the results of the test.

• Project Completed = 50%

• 50% of your grade depends on how much weight your project can hold up compared to others.

• Top scores get +50%, and those following get +40%, +30%, or +20%.

NOTES:

CATEGORIES: Cardboard Tubes, Dead Lift, Materials Strength, Weight

MISSION: Rubber Match

BRIEF:
You and your team have been selected to make the strongest chain possible from a handful of rubber bands!

MISSION RULES:

1. You will design a rubber band chain 24 inches long, and your chain must be created from only rubber bands. Your teacher will decide on your maximum number of rubber bands you may use in your project.

2. Your chain must attach at both ends to test its weight allowance.

3. Your teacher will have a container to attach to one end of your chain and a pole, rope loop, or metal eyelet to attached to the other .

4. You must work with only 1 partner, or alone. Teams may not be made up of more than 2 people.

QUIZ GRADE:

Research and design on chains.

- An explanation of your theory/strategy 25%

- A labeled sketch of your idea 25%

- Conclusions and reflections based on your results 50%

TEST GRADE:

Your completed design and the results of the test.

- Project Completed = 50%

- 50% of your grade depends on how much weight your project holds.

- NOTE: The best project gets an automatic 100%.

NOTES:

CATEGORIES: Chains, Materials Strength, Rubber Bands, Weight

MISSION: Strong As Aluminum

BRIEF: You and your team have been selected to make the strongest chain possible from tin foil!

MISSION RULES:

1. You will design a paper chain at least 12 inches long, and your chain must be created only from tin foil. There may be no other materials, and your teacher will decide how much tin foil each team gets.

2. Your chain must attach at both ends to test its weight allowance.

3. Your teacher will have a container to attach to one end of your chain and a pole, rope loop, or metal eyelet to attached to the other.

4. You must work with only 1 partner, or alone. Teams may not be made up of more than 2 people.

QUIZ GRADE:

Research and design on chains.

• An explanation of your theory/strategy 25%

• A labeled sketch of your idea 25%

• Conclusions and reflections based on your results 50%

TEST GRADE:

Your completed design and the results of the test.

• Project Completed = 50%

• 50% of your grade depends on how much weight your project holds.

• NOTE: The best project gets an automatic 100%.

NOTES:

CATEGORIES: Chains, Foil, Materials Strength, Weight

MISSION: Tear the Trampoline I - Plastic Wrap

BRIEF: You and your team have been selected to make the strongest possible trampoline with plastic wrap!

MISSION RULES:

1. You will design a device that holds a 12x12 inch piece of plastic wrap.

2. Your device will be set up over a gap between tables.

3. Your device will be some sort of frame designed to hold the plastic wrap sheet without tearing it as weight is added.

4. If the plastic wrap tears or is ripped out of your device, no more weight will be added, and the test will be over.

5. You will work with one to two partners. Teams may not be of more than 3 people.

6. You may use any approved materials for your product, including: popsicle sticks, glue, toothpicks, paper, tape, card stock, etc...

7. Your device must be free-standing and movable. It cannot be attached to any surface.

TEACHER'S NOTES: Suggested weights are: marbles, pennies, or graduated weights. It might be a good idea to have a bucket to get the weights when they fall through the device.

QUIZ GRADE:

Create a blueprint design for your ideas

- Sketch 25%

- Sketch is labeled 25%

- Explanation of strategies 25%

- Conclusions and reflections based on your results 25%

TEST GRADE:

Your completed design and the results of the test.

- Project Completed = 50%

- 50% of your grade depends on how much weight your project can hold up compared to others.

- Top scores get +50%, and those following get +40%, +30%, or +20%.

NOTES:

CATEGORIES: Materials Strength, Plastic Wrap, Trampolines, Weight

MISSION: Tear the Trampoline II - Wax Paper

BRIEF: You and your team have been selected to make the strongest possible trampoline with wax paper!

MISSION RULES:

1. You will design a device that holds a 12x12 inch piece of wax paper.

2. Your device will be set up over a gap between tables.

3. Your device will be some sort of frame designed to hold the wax paper sheet without tearing it as weight is added.

4. If the wax paper tears or is ripped out of your device, no more weight will be added, and the test will be over.

5. You will work with one to two partners. Teams may not be of more than 3 people.

6. You may use any approved materials for your product, including: popsicle sticks, glue, toothpicks, paper, tape, card stock, etc...

7. Your device must be free-standing and movable. It cannot be attached to any surface.

TEACHER'S NOTES: Suggested weights are: marbles, pennies, or graduated weights. It might be a good idea to have a bucket to get the weights when they fall through the device.

QUIZ GRADE:

Create a blueprint design for your ideas

- Sketch 25%

- Sketch is labeled 25%

- Explanation of strategies 25%

- Conclusions and reflections based on your results 25%

TEST GRADE:

Your completed design and the results of the test.

- Project Completed = 50%

- 50% of your grade depends on how much weight your project can hold up compared to others.

- Top scores get +50%, and those following get +40%, +30%, or +20%.

NOTES:

CATEGORIES: Materials Strength, Trampolines, Wax Paper, Weight

MISSION: Tear the Trampoline III - Tissue

BRIEF: You and your team have been selected to make the strongest possible trampoline with tissue!

MISSION RULES:

1. You will design a device that holds a piece of tissue paper, kleenex, or paper towel, as determined by your teacher.

2. Your device will be set up over a gap between tables.

3. Your device will be some sort of frame designed to hold the tissue without tearing it as weight is added.

4. If the tissue tears or is ripped out of your device, no more weight will be added, and the test will be over.

5. You will work with one to two partners. Teams may not be of more than 3 people.

6. You may use any approved materials for your product, including: popsicle sticks, glue, toothpicks, paper, tape, card stock, etc...

7. Your device must be free-standing and movable. It cannot be attached to any surface.

TEACHER'S NOTES: Suggested weights are: marbles, pennies, or graduated weights. It might be a good idea to have a bucket to get the weights when they fall through the device.

QUIZ GRADE:

Create a blueprint design for your ideas

- Sketch 25%

- Sketch is labeled 25%

- Explanation of strategies 25%

- Conclusions and reflections based on your results 25%

TEST GRADE:

Your completed design and the results of the test.

- Project Completed = 50%

- 50% of your grade depends on how much weight your project can hold up compared to others.

- Top scores get +50%, and those following get +40%, +30%, or +20%.

NOTES:

CATEGORIES: Materials Strength, Tissue, Trampolines, Weight

MISSION: Tear the Trampoline IV - Paper

BRIEF:
You and your team have been selected to make the strongest possible trampoline with paper!

MISSION RULES:

1. You will design a device that holds a piece of paper.

2. Your device will be set up over a gap between tables.

3. Your device will be some sort of frame designed to hold the paper without tearing it as weight is added.

4. If the paper tears or is ripped out of your device, no more weight will be added, and the test will be over.

5. You will work with one to two partners. Teams may not be of more than 3 people.

6. You may use any approved materials for your product, including: popsicle sticks, glue, toothpicks, paper, tape, card stock, etc...

7. Your device must be free-standing and movable. It cannot be attached to any surface.

TEACHER'S NOTES: Suggested weights are: marbles, pennies, or graduated weights. It might be a good idea to have a bucket to get the weights when they fall through the device.

QUIZ GRADE:

Create a blueprint design for your ideas

- Sketch 25%

- Sketch is labeled 25%

- Explanation of strategies 25%

- Conclusions and reflections based on your results 25%

TEST GRADE:

Your completed design and the results of the test.

- Project Completed = 50%

- 50% of your grade depends on how much weight your project can hold up compared to others.

- Top scores get +50%, and those following get +40%, +30%, or +20%.

NOTES:

CATEGORIES: Materials Strength, Paper, Trampolines, Weight

MISSION: Weakest Link

BRIEF: You and your team have been selected to make the strongest chain possible from paper clips!

MISSION RULES:

1. You will design a paper clip chain at least 12 inches long, and your chain must be created from ONL Y paper clips. Your teacher will decide on your maximum number of paperclips.

2. Your chain must attach at both ends to eyelets or metal loops for testing its weight allowance.

3. One eyelet or metal loop will be secured to a board or a surface. It will not move. The other will hang at the other end of the chain and attach to a bucket or container. Weight will be added to the container slowly. The test is over when your chain gives out.

4. You must work with only 1 partner, or alone. Teams may not be made up of more than 2 people.

QUIZ GRADE:

A blueprint design of your idea

- Sketch 25%

- Sketch is labeled 25%

- Explanation of strategy 25%

- Conclusions and reflections based on your results 25%

TEST GRADE:

Your completed design and the results of the test.

- Project Completed = 50%

- 50% of your grade depends on how much weight your project holds.

- NOTE: The best project gets an automatic 100%.

NOTES:

CATEGORIES: Chains, Materials Strength, Paper Clips, Weight

MISSION: Webbed Up

BRIEF: You and your team have been selected to make a web device that can hold as much weight as possible without breaking.

MISSION RULES:

1. You will design a device using only string or yarn.

2. No single piece of yarn or string may measure more than 12 inches.

3. You may only tie knots to attach the strings. You may not use glue or other adhesives.

4. The device must only be attached at 4-6 anchor points provided by your instructor.

5. You will work with one or two partners. Teams may be of no more than 3 people.

6. Once your device is tied to the anchors, weight will be added gradually to determine how much it can hold before dropping weights or ripping.

TEACHER'S NOTES: C-Clamps, nails in boards, or other anchors should be provided. 4 is a minimum, while 6 (3 per side or in a hex pattern) are suggested.

Suggested weights are: old textbooks, bags of potatoes or similarly-weighted objects. Weights should be large enough to catch in the nets without falling through gaps.

QUIZ GRADE:

A blueprint design of your idea

- Sketch 25%

- Sketch is labeled 25%

- Explanation of strategy 25%

- Conclusions and reflections based on your results 25%

TEST GRADE:

Your completed design and the results of the test.

- Project Completed = 50%

- 50% of your grade depends on how much weight your project holds.

- NOTE: The best project gets an automatic 100%.

NOTES:

CATEGORIES: Materials Strength, String, Weight, Yarn

150 STEM Labs

SCIENCE * TECHNOLOGY * ENGINEERING * MATH

Task Completion Projects

MISSION: Cable Cars I - Weight

BRIEF: You and your team have been selected to make a cable car that can hold as much weight as possible.

MISSION RULES:

1. You will design a cable car that is attached to a string with a sliding plastic straw. The cable car must attach to the string, but it does not have to slide.

2. Your cable car must be built from a plastic straw, tape or glue, notecards, and other teacher-approved materials.

3. You will work with a one or two partners. Teams may not be of more than 3 people.

4. Your device must have a cup or receiver for weights to be placed in. If the cable car breaks or if your cup is full, no more weight will be added and the test will be over.

TEACHER'S NOTES: Suggested weights are pennies, paper clips, sand, or graduated weights.

For the line, heavy fishing line is suggested, anchored and tied on one end. The other end should be tied and retied for each test. Some sort of eye-bolt or carabiner might help. You might need to put all the straws on the line FIRST, and then tie it. Cable cars would then be designed in place.

QUIZ GRADE:

Create a blueprint designs for your ideas

- Sketch 25%

- Sketch is labeled 25%

- Explanation of strategies 25%

- Conclusions and reflections based on your results 25%

TEST GRADE:

Your completed design and the results of the test.

- Project Completed = 50%

50% of your grade depends on how much weight your project holds compared to others.

- Other places = +10-20%

- Third Place = +30%

- Second Place = +40%

- Holds the Most Weight = +50%

NOTES:

CATEGORIES: Cable Cars, Materials Strength, Scavengers, String, Weight

MISSION: Cable Cars II - Water

BRIEF: You and your team have been selected to make a cable car that can hold as much water as possible.

MISSION RULES:

1. You will design a cable car that is attached to a string with a sliding plastic straw. The cable car must attach to the string, but it does not have to slide.

2. Your cable car must be built from a plastic straw, plastic, styrofoam, foil, tape or glue, notecards, and other teacher-approved materials.

3. You will work with a one or two partners. Teams may not be of more than 3 people.

4. Your device must have a cup or receiver for water to be placed in. If the cable car breaks, leaks, or your cup is full, no more water will be added and the test will be over.

TEACHER'S NOTES: Keep a bucket under the test projects or test outside.

For the line, heavy fishing line is suggested, anchored and tied on one end. The other end should be tied and retied for each test. Some sort of eye-bolt or carabiner might help. You might need to put all the straws on the line FIRST, and then tie it. Cable cars would then be designed in place.

QUIZ GRADE:

Create a blueprint designs for your ideas

- Sketch 25%

- Sketch is labeled 25%

- Explanation of strategies 25%

- Conclusions and reflections based on your results 25%

TEST GRADE:

Your completed design and the results of the test.

- Project Completed = 50%

50% of your grade depends on how much water your project holds compared to others.

- Other places = +10-20%

- Third Place = +30%

- Second Place = +40%

- Holds the Most Water = +50%

NOTES:

CATEGORIES: Cable Cars, Capacity, Materials Strength, Scavengers, String, Water

MISSION: Coin Collection

BRIEF: You and your team have been selected to make a device that sorts 3 types of coins into different containers.

MISSION RULES:

1. You will design a sorting device.

2. Your device may be any shape or size, but it must sort the 3 types of coins into 3 different containers, and they must be properly sorted.

3. You will work with one or two partners. Teams may not be of more than 3 people.

4. You may use any approved materials you can find at school or at home, including paper clips, pipe cleaners, plastic straws, notecards, tape...

TEACHER'S NOTES: Use 10 coins of assorted varieties. Quarters, Nickels, and Pennies are suggested for the variety of sizes and thicknesses. If this proves to be too hard, switch it to 2 kinds of coins, probably pennies and quarters.

QUIZ GRADE:

Create a blueprint design for your ideas

• Sketch 25%

• Sketch is labeled 25%

• Explanation of strategies 25%

• Conclusions and reflections based on your results 25%

TEST GRADE:

Your completed design and the results of the test.

• Project Completed = 50%

• 50% of your grade depends on the success of your sorting. Each coin is worth 5%, and you must sort 10 coins.

NOTES:

CATEGORIES: Coins, Scavengers, Sorting

MISSION: Eggs in a Basket

BRIEF: You and your team have been selected to make a device that allows a ping pong ball in a cup to land on the floor safely without the ball falling out of the cup.

MISSION RULES:

1. You will design a device that allows a ping pong ball in a cup to land on the floor safely without the ball falling out of the cup.

2. Your device may be of any dimensions under 18 inches in any one direction.

3. You may build your device from any approved materials found at school or at home.

4. Your cup may not be covered at all. Only hooks/line may be attached around the rim of the cup to allow it to be hooked to your project.

5. You will work with one or two partners. Teams may be of no more than 3 people.

6. Up to three tests will be made. Projects will be dropped from a height of no less than 6 feet.

TEACHER'S NOTES: Standing on top of a chair/table or step ladder is advised. Dropping from farther up may give certain projects a chance to deploy.

Substitute a marble if ping pong balls are not available.

QUIZ GRADE:

A blueprint design of your idea

- Sketch 25%

- Sketch is labeled 25%

- Explanation of strategy 25%

- Conclusions and reflections based on your results 25%

TEST GRADE:

Your completed design and the results of the test.

- Project Completed = 50%

- 50% of your grade depends on test results

  - Survive 3 landings = 50%

  - Survive 2 landings = 35%

  - Survive 1 landing = 20%

  - Survive 0 landings = 0%

NOTES:

CATEGORIES: Cups, Ping Pong Balls, Scavengers, Survival

MISSION: Homemade Orchestra I - Percussion

BRIEF: You and your team have been selected to make a custom percussion instrument from scavenged materials.

MISSION RULES:

1. You will research percussion instruments to get ideas for a concept for your design.

2. Your design may be of any reasonable dimensions.

3. You will work with two or three partners. Teams may not be of more than 4 people.

4. You may use any approved scavenged materials from home and school to build your device.

5. Your device must have a minimum of 3 different audible tones. It can be played by hand or with some sort of stick.

6. Your team must compose and perform a song at least 30 seconds in length with your instrument. Practice makes perfect!

QUIZ GRADE:

A research paper on percussion instruments.

- 2-4 pictures of percussion instruments 25%

- A concept idea for your percussion instruments 50%

- Conclusions and reflections based on your results 25%

TEST GRADE:

Your completed design and the results of the test.

- Project Completed = 60%

- 40% of your grade depends on your instrument's performance. If the instrument breaks during the performance, it will cost points.

- Writing the music on music sheet notes is a BONUS!

NOTES:

CATEGORIES: Music, Scavengers, Sound, Time

MISSION: Homemade Orchestra II - Strings

BRIEF: You and your team have been selected to make a custom stringed instrument from scavenged materials.

MISSION RULES:

1. You will research stringed instruments to get ideas for a concept for your design.

2. Your design may be of any reasonable dimensions.

3. You will work with two or three partners. Teams may not be of more than 4 people.

4. You may use any approved scavenged materials from home and school to build your device.

5. Your device must have a minimum of 3 strings that give different audible tones. It can be plucked, strummed, or it may require a custom-made pick or bow to play it.

6. Your team must compose and perform a song at least 30 seconds in length with your instrument. Practice makes perfect!

QUIZ GRADE:

A research paper on stringed instruments.

- 2-4 pictures of stringed instruments 25%

- A concept idea for your stringed instruments 50%

- Conclusions and reflections based on your results 25%

TEST GRADE:

Your completed design and the results of the test.

- Project Completed = 60%

- 40% of your grade depends on your instrument's performance. If the instrument breaks during the performance, it will cost points.

- Writing the music on music sheet notes is a BONUS!

NOTES:

CATEGORIES: Music, Scavengers, Sound, String, Time

MISSION: Homemade Orchestra III - Wind

BRIEF:

You and your team have been selected to make a custom wind (brass or woodwind) instrument from scavenged materials.

MISSION RULES:

1. You will research wind instruments to get ideas for a concept for your design.

2. Your design may be of any reasonable dimensions.

3. You will work with one or two partners. Teams may not be of more than 3 people.

4. You may use any approved scavenged materials from home and school to build your device.

5. Your device must have a minimum of 3 different audible tones.

6. Your team must compose and perform a song at least 30 seconds in length with your instrument. Practice makes perfect!

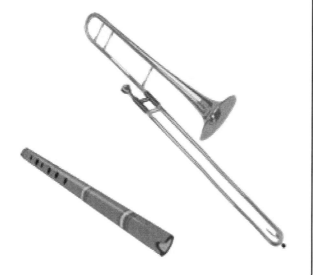

QUIZ GRADE:

A research paper on wind instruments (including brass and woodwind).

- 2-4 pictures of wind instruments 25%

- A concept idea for your wind instruments 50%

- Conclusions and reflections based on your results 25%

TEST GRADE:

Your completed design and the results of the test.

- Project Completed = 60%

- 40% of your grade depends on your instrument's performance. If the instrument breaks during the performance, it will cost points.

- Writing the music on music sheet notes is a BONUS!

NOTES:

CATEGORIES: Music, Scavengers, Sound, Time, Wind

MISSION: Homemade Orchestra IV - Full Orchestra

BRIEF: You and your team have been selected to make at least 3 instruments of different families and compose a tune together.

MISSION RULES:

1. You will research instruments to get ideas for a concept for your designs.

2. Your design may be of any reasonable dimensions.

3. You will work with three to five partners. Teams may not be of more than 6 people.

4. You may use any approved scavenged materials from home and school to build your device.

5. Each device must have a minimum of 3 different audible tones.

6. Your team must compose and perform a song at least 60 seconds in length with your instruments. Practice makes perfect!

QUIZ GRADE:

A research paper on typical instruments in an orchestra.

- 5-10 pictures of wind instruments 25%

- Concept ideas for each of your instruments 50%

- Conclusions and reflections based on your results 25%

TEST GRADE:

Your completed design and the results of the test.

- Project Completed = 60%

- 40% of your grade depends on your band's performance. If the instruments break during the performance, it will cost points.

- Writing the music on music sheet notes is a BONUS!

NOTES:

CATEGORIES: Music, Scavengers, Sound, Time, Wind

MISSION: Let's Get Cooking

BRIEF:

You and your team have been selected to make a solar oven.

MISSION RULES:

1. You will design an oven.

2. Your device may be any dimensions less than 12 inches in any one direction.

3. You will work with one or two partners. Teams may not be of more than 3 people.

4. You may use only foil and other approved materials found at home or school to build your oven.

5. The oven must melt pieces of chocolate as completely and quickly as possible. A paperclip or probe will be used to determine how melted the chocolate is, if there is any question.

TEACHER'S NOTES: small pieces of milk chocolate (3 or more) should be spread around the baking surface in full sun to determine which ovens work best.

QUIZ GRADE:

A blueprint design of your idea

- Sketch 25%

- Sketch is labeled 25%

- Explanation of strategy 25%

- Conclusions and reflections based on your results 25%

TEST GRADE:

Your completed design and the results of the test.

- Project Completed = 50%

- 50% of your grade depends on how fast and completely your project melts the chocolate. Better projects score more points.

NOTES:

CATEGORIES: Chocolate, Foil, Heat, Melting, Scavengers, Sunlight, Time

MISSION: Lost Your Marbles

BRIEF: You and your team have been selected to make a device that sorts marbles and ping pong balls into different containers.

MISSION RULES:

1. You will design a sorting device.

2. Your device may be any shape or size, but it must sort the 2 types of balls into 2 different containers, and they must be properly sorted.

3. You will work with one or two partners. Teams may not be of more than 3 people.

4. You may use any approved materials you can find at school or at home, including paper clips, pipe cleaners, plastic straws, notecards, tape...

TEACHER'S NOTES: Use 5 total marbles and ping pong balls.

QUIZ GRADE:

Create a blueprint design for your ideas

- Sketch 25%

- Sketch is labeled 25%

- Explanation of strategies 25%

- Conclusions and reflections based on your results 25%

TEST GRADE:

Your completed design and the results of the test.

- Project Completed = 50%

- 50% of your grade depends on the success of your sorting. Each ball is worth 10%, and you must sort 5 balls or marbles.

NOTES:

CATEGORIES: Marbles, Ping Pong Balls, Scavengers, Sorting

MISSION: Pin the Tail on the Balloon

BRIEF: You and your team have been selected to make a device that can pop balloons.

MISSION RULES:

1. You will design a device that will pop balloons.

2. You may use any approved materials that you find in the classroom or at home.

3. You may not simply drop a weight on the balloon. There must be a device/lever/ mechanism that creates an action that pops the balloon.

4. You will work with one to two partners. Teams may not be of more than 3 people.

5. 3 tests will be made. Your teacher may decide on a best of 3, an average score, or some composite score.

TEACHER'S NOTES: Try balloons of varying levels of inflation. Less inflated balloons are often harder to pop. You will want to carefully monitor what potentially dangerous objects are used to pop the balloons. :)

QUIZ GRADE:

Create a blueprint design for your ideas

- Sketch 25%

- Sketch is labeled 25%

- Explanation of strategies 25%

- Conclusions and reflections based on your results 25%

TEST GRADE:

Your completed design and the results of the test.

- Project Completed = 25%

- 75% of your grade depends on the results of your trials. Each successful test is worth 25%.

NOTES:

CATEGORIES: Balloons, Crashes, Scavengers

MISSION: Nice Cold Drink

BRIEF:
You and your team have been selected to make a device that cleans water.

MISSION RULES:

1. You will design a device to clean water.

2. Your device may be any size, but it must have a place to put dirty water into the device and a place for clean water to collect.

3. You will work with one or two partners. Teams may not be of more than 3 people.

4. You may use any approved materials you can find at school or at home, including paper clips, pipe cleaners, plastic straws, notecards, tape...

TEACHER'S NOTES: It is suggested to have a variety of pollutants in the water of varying sizes. Dirt, sawdust, gravel, and even food coloring can be added. Better filters will catch more sizes and types of debris.

QUIZ GRADE:

Research and design of a water filtration device.

- A paragraph on water filtration 25%

- A concept idea for your filtration device sketched and explained 50%

- Conclusions and reflections based on your results 25%

TEST GRADE:

Your completed design and the results of the test.

- Project Completed = 50%

- 50% of your grade depends on if your project actually works. Cleaner = better grade.

- NOTE: The best project gets an automatic 100%.

NOTES:

CATEGORIES: Scavengers, Sorting, Water

MISSION: Running Uphill

BRIEF: You and your team have been selected to make a device that will move water from one cup up to another higher cup.

MISSION RULES:

1. Your design may be of any dimensions less than 18 inches in any one direction.

2. Your device must incorporate a cup a the beginning and the end. Water must be moved from one cup up to the other.

3. You will work with two or three partners. Teams may not be of more than 4 people.

4. You may use any approved scavenged materials from home and school to build your device.

5. There is a 2 minute time limit. Any water not moved at that point is considered lost.

QUIZ GRADE:

A research paper on pumps and water movement devices

- 2-4 pictures of devices 25%

- A labeled blueprint concept idea for your devices 50%

- Conclusions and reflections based on your results 25%

TEST GRADE:

Your completed design and the results of the test.

- Project Completed = 50%

- 50% of your grade depends on how fast you can move the water from one cup to the other. Faster projects score better.

NOTES:

CATEGORIES: Cups, Scavengers, Time, Water

MISSION: Versus I - The Takedown

BRIEF: One team must design a tower. The other team must knock it down.

MISSION RULES:

1. Each team will consist of 2-4 members.

2. A-Teams will design towers. B-Teams will design machines to knock them down.

3. Each team may design a device from any approved materials found at home or at school.

4. A-Teams and B-Teams should design and build their devices separately.

5. Devices should not be more than 18 inches in any one dimension, and no more than 12 inches in the other dimensions.

6. When completed, A-Teams' towers must survive 5 attacks from B-Teams' devices.

TEACHER'S NOTES: Choosing A-Teams and B-Teams may be done randomly.

You may also make all students be A's or B's, while you take the other job.

QUIZ GRADE:

A blueprint design of your idea

- Sketch 25%

- Sketch is labeled 25%

- Explanation of strategy 25%

- Conclusions and reflections based on your results 25%

TEST GRADE:

Your completed design and the results of the test.

- Project Completed = 50%

- 50% of your grade depends on how much well your project does.

  - A-TEAMS: Project survives with no damage 50%, minimal damage 40%, some damage 30%, completely wrecked, 20%

  - B-TEAMS: Project inflicts complete destruction 50%, mostly complete destruction 40%, some destruction 30%, minimal destruction 20%

CATEGORIES: Crashes, Scavengers, Survival, Throwers, Towers, Versus

MISSION: Versus II - The Splashdown

BRIEF:

One team must design a boat. The other team must destroy or sink it.

MISSION RULES:

1. Each team will consist of 2-4 members.

2. A-Teams will design boats. B-Teams will design machines to sink or break them.

3. Each team may design a device from any approved materials found at home or at school.

4. A-Teams and B-Teams should design and build their devices separately.

5. A-Team boats should not be more than 9 inches in any dimension. B-Team devices Devices should not be more than 12 inches in any one dimension.

6. When completed, A-Teams' boats must survive 5 attacks from B-Teams' devices.

TEACHER'S NOTES: Choosing A-Teams and B-Teams may be done randomly.

You may also make all students be A's or B's, while you take the other job.

QUIZ GRADE:

A blueprint design of your idea

- Sketch 25%

- Sketch is labeled 25%

- Explanation of strategy 25%

- Conclusions and reflections based on your results 25%

TEST GRADE:

Your completed design and the results of the test.

- Project Completed = 50%

- 50% of your grade depends on how much well your project does.

  - A-TEAMS: Project survives with no damage 50%, minimal damage 40%, some damage 30%, completely wrecked, 20%

  - B-TEAMS: Project inflicts complete destruction 50%, mostly complete destruction 40%, some destruction 30%, minimal destruction 20%

CATEGORIES: Boats, Crashes, Scavengers, Survival, Throwers, Versus, Water

MISSION: Versus III - The Crashdown

BRIEF: One team must design a catching device. The other team must throw a golf ball through it or knock it down.

MISSION RULES:

1. Each team will consist of 2-4 members.

2. A-Teams will design a catching device that gathers golf balls into a cup. B-Teams will design machines to throw golf balls at those devices.

3. Each team may design a device from any approved materials found at home or at school.

4. A-Teams and B-Teams should design and build their devices separately.

5. Either type of device should not be more than 12 inches in any one dimension.

6. When completed, A-Teams' nets must survive and catch as many of the 5 golf ball throws from B-Teams' devices as possible. A ball that misses the device entirely does not count.

7. Success for the catching device is capturing and holding the golf ball in a tank or trap without breaking or dropping the ball.

TEACHER'S NOTES: Choosing A-Teams and B-Teams may be done randomly.

You may also make all students be A's or B's, while you take the other job.

QUIZ GRADE:

A blueprint design of your idea

- Sketch 25%

- Sketch is labeled 25%

- Explanation of strategy 25%

- Conclusions and reflections based on your results 25%

TEST GRADE:

Your completed design and the results of the test.

- Project Completed = 50%

- 50% of your grade depends on how much well your project does.

  - A-TEAMS: Project catches all 5 balls without falling over or getting broken 50%, 4 balls 40%, 3 balls 30%, 2 balls 20%, 1 ball 10%, no balls 0%

  - B-TEAMS: Other project is unable to catch and stop all 5 balls 50%, drops 4 balls 40%, drops 3 balls 30%, drops 2 balls 20%, drops only 1 ball 10%, catches all the balls 0%

CATEGORIES: Crashes, Cups, Golf Balls, Scavengers, Survival, Throwers, Versus

MISSION: Versus IV - The Rundown

BRIEF: One team must design a catching device. The other team must roll a car through it or knock it down.

MISSION RULES:

1. Each team will consist of 2-4 members.

2. A-Teams will design a catching device that can stop a car without being broken or knocked over. B-Teams will design cars to roll at those devices and crash through them.

3. A-Teams should build their devices of packing peanuts and toothpicks.

4. B-Teams should design a car from scavenged materials that can move under its own power.

5. A-Teams and B-Teams should design and build their devices separately.

6. Either type of device should not be more than 12 inches in any one dimension.

7. When completed, A-Teams' nets must survive and catch B-Teams' cars as many times as possible out of 5 trials.

8. Success for the catching device is stopping or holding the car without breaking or being knocked over. If the car misses entirely, redo the test run.

TEACHER'S NOTES: Choosing A-Teams and B-Teams may be done randomly.

You may also make all students be A's or B's, while you take the other job.

QUIZ GRADE:

A blueprint design of your idea

- Sketch 25%

- Sketch is labeled 25%

- Explanation of strategy 25%

- Conclusions and reflections based on your results 25%

TEST GRADE:

Your completed design and the results of the test.

- Project Completed = 50%

- 50% of your grade depends on how much well your project does.

  - A-TEAMS: Project catches all 5 cars without falling over or getting broken 50%, 4 cars 40%, 3 cars 30%, 2 cars 20%, 1 car 10%, no balls 0%

  - B-TEAMS: Other project is broken or fails to stop the car 5 times 50%, 4 times 40%, 3 times 30%, 2 times 20%, 1 time 10%, no times 0%.

CATEGORIES: Cars, Crashes, Packing Peanuts, Toothpicks, Scavengers, Survival, Versus

MISSION: Washboard

BRIEF: You and your team have been selected to make a device that will squeeze as much water as possible out of a wet sponge.

MISSION RULES:

1. Your design may be of any dimensions less than 18 inches in any one direction.

2. There must be a place to put the wet sponge in your machine, where it can be crushed or squeezed.

3. Your device must incorporate a catch basin to collect water that is squeezed.

4. You will work with two or three partners. Teams may not be of more than 4 people.

5. You may use any approved scavenged materials from home and school to build your device.

6. There must be an activation lever or something that starts your machine.

7. There is a 1 minute time limit. At that time, the test is considered finished.

QUIZ GRADE:

A blueprint design of your idea

- Sketch 25%

- Sketch is labeled 25%

- Explanation of strategy 25%

- Conclusions and reflections based on your results 25%

TEST GRADE:

Your completed design and the results of the test.

- Project Completed = 50%

- 50% of your grade depends on how dry you can get the sponge. Better projects will score higher.

NOTES:

CATEGORIES: Scavengers, Sponges, Time, Water

MISSION: Water Delivery

BRIEF: You and your team have been selected to make a device that will move water from one cup to another.

MISSION RULES:

1. Your design may be of any dimensions less than 18 inches in any one direction.

2. Your device must incorporate a cup a the beginning and the end. Water must be moved from one cup to the other.

3. The first cup cannot simply be poured/ dumped by hand, either.

4. You will work with two or three partners. Teams may not be of more than 4 people.

5. You may use any approved scavenged materials from home and school to build your device.

6. There is a 2 minute time limit. Any water not moved at that point is considered lost.

QUIZ GRADE:

A research paper on pumps and water movement devices

- 2-4 pictures of devices 25%

- A labeled blueprint concept idea for your devices 50%

- Conclusions and reflections based on your results 25%

TEST GRADE:

Your completed design and the results of the test.

- Project Completed = 50%

- 50% of your grade depends on how fast you can move the water from one cup to the other. Faster projects score better.

NOTES:

CATEGORIES: Cups, Scavengers, Time, Water

MISSION: What's for Breakfast?

BRIEF: You and your team have been selected to make a device that sorts different shapes and sizes of breakfast cereal into different containers.

MISSION RULES:

1. You will design a sorting device.

2. Your device may be any shape or size, but it must sort the 2 or more types of cereal into different containers, and they must be properly sorted.

3. You will work with one or two partners. Teams may not be of more than 3 people.

4. You may use any approved materials you can find at school or at home, including paper clips, pipe cleaners, plastic straws, notecards, tape...

TEACHER'S NOTES: Use 2 or more different shapes and styles of cereal. Square biscuits, balls, and round rings might be good types to try.

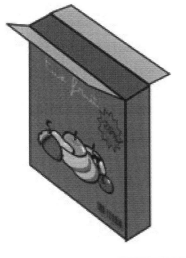

QUIZ GRADE:

Create a blueprint design for your ideas

- Sketch 25%

- Sketch is labeled 25%

- Explanation of strategies 25%

- Conclusions and reflections based on your results 25%

TEST GRADE:

Your completed design and the results of the test.

- Project Completed = 50%

- 50% of your grade depends on the success of your sorting. Each ball is worth 10%, and you must sort 5 balls or marbles.

NOTES:

CATEGORIES: Cereal, Scavengers, Sorting

150 STEM Labs

SCIENCE * TECHNOLOGY * ENGINEERING * MATH

Throwers Projects

MISSION: 3-2-1 Launch

BRIEF: You and your team have been selected to make a device to throw a ping pong ball.

MISSION RULES:

1. You will design a throwing device.

2. Your device must be no longer than 18 inches, no taller than 18 inches, and no wider than 12 inches when assembled and stationed at the throwing line.

3. You will work with two or three partners. Teams may not be of more than 4 people.

4. You must only use paper, glue, tape, rubber bands, paperclips, pencils, or other approved office supplies for your project.

5. The device must have some cup or place to put the ping pong ball. The device will then be manipulated and the attempt measured.

QUIZ GRADE:

Create a blueprint design for your ideas

- Sketch 25%

- Sketch is labeled 25%

- Explanation of strategies 25%

- Conclusions and reflections based on your results 25%

TEST GRADE:

Your completed design and the results of the test.

- Project Completed = 50%

- 50% of your grade depends on how far your project throws compared to the other group's projects. The projects that do best will get more points.

- NOTE: There is a -5% penalty for every 1/2 inch your project is out of the size specifications.

NOTES:

CATEGORIES: Distance, Ping Pong Balls, Rubber Bands, Throwers

MISSION: Blowguns

BRIEF: You and your team have been selected to make a projectile that can be shot as far as possible with a plastic straw.

MISSION RULES:

1. You will design a set of **harmless** projectiles that shoot as far as possible by blowing into a plastic straw.

2. Your projectiles/devices may be of any dimensions, but they must have a receiver that allows it to fit in or around the end of the straw.

3. You may build your device from any approved materials found at school or at home.

4. You will work with one or two partners. Teams may be of no more than 3 people.

5. Up to three tests will be made. Distance will be measured at the first contact with the ground.

TEACHER'S NOTES: After 3 tests, honors can be given for the best average and/or the highest single shot.

PLEASE NOTE: Take precautions not to inhale any projects or to hit anyone in the eyes.

QUIZ GRADE:

A blueprint design of your idea

- Sketch 25%

- Sketch is labeled 25%

- Explanation of strategy 25%

- Conclusions and reflections based on your results 25%

TEST GRADE:

Your completed design and the results of the test.

- Project Completed = 50%

- 50% of your grade depends on how far your project sends the projectile.

- NOTE: The best project gets an automatic 100%.

NOTES:

CATEGORIES: Air, Distance, Plastic Straws, Throwers

MISSION: Bottle Blasters

BRIEF: You and your team have been selected to make a device that flies as far as possible by squeezing an empty plastic bottle.

MISSION RULES:

1. You will design a **harmless** flying device that is propelled by squeezing (or stomping on) an empty plastic bottle.

2. Your devices may be of any dimensions, but they must have a receiver that allows it to fit in or around the cap end of the plastic bottle.

3. You may build your device from any approved materials found at school or at home.

4. You will work with one or two partners. Teams may be of no more than 3 people.

5. Up to three tests will be made. Distance will be measured at the first contact with the ground.

TEACHER'S NOTES: After 3 tests, honors can be given for the best average and/or the highest single shot.

Please take eye safety precautions.

QUIZ GRADE:

A blueprint design of your idea

- Sketch 25%

- Sketch is labeled 25%

- Explanation of strategy 25%

- Conclusions and reflections based on your results 25%

TEST GRADE:

Your completed design and the results of the test.

- Project Completed = 50%

- 50% of your grade depends on how far your project flies

- NOTE: The best project gets an automatic 100%.

NOTES:

CATEGORIES: Air, Bottles, Distance, Scavengers, Throwers

MISSION: Crossbows

BRIEF: You and your team have been selected to make a device that shoots an arrow as far as possible.

MISSION RULES:

1. You will design a device that shoots a blunt or **harmless** arrow as far as possible using rubber bands as your major method of propulsion.

2. Your device may be of any dimensions under 9 inches in any one direction.

3. You may build your device from any approved materials found at school or at home.

4. The device must be freestanding and not attached to any surface.

5. You will work with one or two partners. Teams may be of no more than 3 people.

6. Up to three tests will be made. Distance will be measured at the first contact with the ground.

TEACHER'S NOTES: After 3 tests, honors can be given for the best average and/or the highest single shot.

You can also test for accuracy.

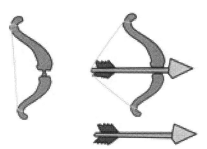

QUIZ GRADE:

A blueprint design of your idea

- Sketch 25%

- Sketch is labeled 25%

- Explanation of strategy 25%

- Conclusions and reflections based on your results 25%

TEST GRADE:

Your completed design and the results of the test.

- Project Completed = 50%

- 50% of your grade depends on how far your project sends the arrow.

- NOTE: The best project gets an automatic 100%.

NOTES:

CATEGORIES: Distance, Rubber Bands, Scavengers, Throwers

MISSION: Discus

BRIEF: You and your team have been selected to make a device that can throw a coin as far as possible.

MISSION RULES:

1. You will design a device that launches a coin as far as possible.

2. Your building materials should consist of paper, card stock, glue, rubber bands, popsicle sticks, and other readily-available materials that your teacher approves.

3. Your device may be no longer than 24 inches in any dimension, and no more than 48 inches total dimensions (l + w + h).

4. You will work with one or two partners. Teams may be of no more than 3 people.

5. Up to three tests will be made. Your teacher will determine how you are scored: longest shot, average, or total.

TEACHER'S NOTES: Some care should be taken during test, so people are not injured by flying coins.

QUIZ GRADE:

A blueprint design of your idea

- Sketch 25%

- Sketch is labeled 25%

- Explanation of strategy 25%

- Conclusions and reflections based on your results 25%

TEST GRADE:

Your completed design and the results of the test.

- Project Completed = 50%

- 50% of your grade depends on how far your coins travels.

- NOTE: The best project gets an automatic 100%.

NOTES:

CATEGORIES: Coins, Distance, Throwers

MISSION: Fire Away!

BRIEF: You and your team have been selected to make a rubber band dart thrower that throws a dart as far as possible!

MISSION RULES:

1. You will design not only a slingshot of rubber bands, but also a paper dart that hooks onto it and can be fired for distance.

2. Your device must not use more than four rubber bands and one piece of paper. Other acceptable materials include a paper clip and a small bit of clear tape.

3. You will work with a partner. Teams may not be of more than 2 people.

4. One partner will hold the slingshot, while the other aims and fires the dart.

5. You will fire your project 3 times and add the total distance fired for a final score.

QUIZ GRADE:

Create a blueprint designs for your ideas

- Sketch 25%

- Sketch is labeled 25%

- Explanation of strategies 25%

- Conclusions and reflections based on your results 25%

TEST GRADE:

Your completed design and the results of the test.

- Project Completed = 50%

- 50% of your grade depends on how far your project fires com- pared to the other group's projects. The projects that do best will get more points.

- NOTE: The project with the longest total fired distance automatically gets 100%

NOTES:

CATEGORIES: Distance, Rubber Bands, Throwers

MISSION: Make it Rain!

BRIEF: You and your team have been selected to make a device to throw a water balloon as far as possible.

MISSION RULES:

1. You will design a throwing device.

2. Your device must be no longer than 18 inches, no taller than 18 inches, and no wider than 12 inches when assembled and stationed at the throwing line.

3. You will work with two or three partners. Teams may not be of more than 4 people.

4. You must only use paper, glue, tape, rubber bands, paperclips, pencils, popsicle sticks, or other approved office supplies for your project.

5. The device must have some cup or place to put the small water balloon. The device will then be manipulated and the attempt measured.

TEACHER'S NOTES: 3 attempts are suggested. A total distance category and/or an average category could be considered for special honors. Mini water balloons are highly suggested, as is doing this outside!

QUIZ GRADE:

Create a blueprint design for your ideas

- Sketch 25%

- Sketch is labeled 25%

- Explanation of strategies 25%

- Conclusions and reflections based on your results 25%

TEST GRADE:

Your completed design and the results of the test.

- Project Completed = 50%

- 50% of your grade depends on how far your project throws compared to the other group's projects. The projects that do best will get more points.

- NOTE: There is a -5% penalty for every 1/2 inch your project is out of the size specifications.

NOTES:

CATEGORIES: Balloons, Distance, Popsicle Sticks, Rubber Bands, Throwers, Water

MISSION: Marshmallows Away!

BRIEF: You and your team have been selected to make a device to throw mini marshmallows as far as possible.

MISSION RULES:

1. You will design a throwing device that is free-standing. It may not be attached to anything or do more than sit on a desk or table surface.

2. The catch is that it must be made primarily from paperclips, along with whatever approved materials you are given.

3. You will work with one or two partners. Teams may not be of more than 3 people.

4. The device must have some place to attach/hold the mini marshmallow so that it can be launched.

QUIZ GRADE:

Create a blueprint design for your ideas

- Sketch 25%

- Sketch is labeled 25%

- Explanation of strategies 25%

- Conclusions and reflections based on your results 25%

TEST GRADE:

- Your completed design and the results of the test.

- Project Completed = 50%

- 50% of your grade depends on how far your project throws compared to the other group's projects. The projects that do best will get more points.

- NOTE: The team that can throw the marshmallow the farthest gets an automatic 100%

NOTES:

CATEGORIES: Distance, Marshmallows, Paper Clips, Throwers

MISSION: Marshmallow Blaster

BRIEF: You and your team have been selected to make a cannon to shoot a marshmallow through a tube as far as possible.

MISSION RULES:

1. You will design a device that launches a marshmallow through a tube as far as possible.

2. Your primary building materials should be a cardboard tube or some pipe, plus whatever scavenged materials you can find to help propel the marshmallow.

3. You will work with one or two partners. Teams may be of no more than 3 people.

4. Up to three tests will be made.

TEACHER'S NOTES: After 3 tests, honors can be given for the best average and/or the longest single shot.

QUIZ GRADE:

A blueprint design of your idea

- Sketch 25%

- Sketch is labeled 25%

- Explanation of strategy 25%

- Conclusions and reflections based on your results 25%

TEST GRADE:

Your completed design and the results of the test.

- Project Completed = 50%

- 50% of your grade depends on how far your project sends the marshmallows.

- NOTE: The best project gets an automatic 100%.

NOTES:

CATEGORIES: Distance, Marshmallows, Scavengers, Throwers

MISSION: Ping Pong Mortar

BRIEF: You and your team have been selected to make a mortar launcher for a ping pong ball.

MISSION RULES:

1. You will design a device that launches a ping pong ball as high into the air as possible.

2. Your primary building materials should be a cardboard tube(s) and some materials to help fling or launch the ping pong ball through the tube and into the air.

3. You will work with one or two partners. Teams may be of no more than 3 people.

4. Up to three tests will be made.

TEACHER'S NOTES: This is often best done outside, where the height of the launch can be measured against how many bricks the ping pong ball flies above, graduated lines marked on the wall, or even meter sticks taped to the wall.

After 3 tests, honors can be given for the best average and/or the highest single shot.

QUIZ GRADE:

A blueprint design of your idea

- Sketch 25%

- Sketch is labeled 25%

- Explanation of strategy 25%

- Conclusions and reflections based on your results 25%

TEST GRADE:

Your completed design and the results of the test.

- Project Completed = 50%

- 50% of your grade depends on how high your project travels.

- NOTE: The best project gets an automatic 100%.

NOTES:

CATEGORIES: Cardboard Tubes, Height, Ping Pong Balls, Throwers

MISSION: Rain Marbles Down on Them!

BRIEF: You and your team have been selected to make a device to throw a marble as far as possible.

MISSION RULES:

1. You will design a throwing device.

2. Your device must be no longer than 18 inches, no taller than 18 inches, and no wider than 12 inches when assembled and stationed at the throwing line.

3. You will work with two or three partners. Teams may not be of more than 4 people.

4. You must only use paper, glue, tape, rubber bands, paperclips, pencils, popsicle sticks, or other approved office supplies for your project.

5. The device must have some cup or place to put the marble. The device will then be manipulated and the attempt measured.

TEACHER'S NOTES: 3 attempts are suggested. A total distance category and/or an average category could be considered for special honors.

QUIZ GRADE:

Create a blueprint design for your ideas

• Sketch 25%

• Sketch is labeled 25%

• Explanation of strategies 25%

• Conclusions and reflections based on your results 25%

TEST GRADE:

Your completed design and the results of the test.

• Project Completed = 50%

• 50% of your grade depends on how far your project throws compared to the other group's projects. The projects that do best will get more points.

• NOTE: There is a -5% penalty for every 1/2 inch your project is out of the size specifications.

NOTES:

CATEGORIES: Distance, Marbles, Popsicle Sticks, Rubber Bands, Throwers

MISSION: Target Practice

BRIEF: You and your team have been selected to make a device that shoots an arrow as accurately as possible.

MISSION RULES:

1. You will design a device that shoots a blunt or **harmless** arrow through a target ring as accurately as possible. Your device should use rubber bands as your major method of propulsion.

2. Your device may be of any dimensions under 9 inches in any one direction.

3. You may build your device from any approved materials found at school or at home.

4. The device must be freestanding and not attached to any surface.

5. You will work with one or two partners. Teams may be of no more than 3 people.

6. At least to three tests will be made from different distances. Success is measured by passing through the target ring.

TEACHER'S NOTES: In testing for accuracy, you can test distance, or test by making progressively smaller target rings to shoot through and keeping the distance the same. 3 attempts at each range are suggested.

QUIZ GRADE:

A blueprint design of your idea

• Sketch 25%

• Sketch is labeled 25%

• Explanation of strategy 25%

• Conclusions and reflections based on your results 25%

TEST GRADE:

Your completed design and the results of the test.

• Project Completed = 50%

• 50% of your grade depends on how accurately your project can put the arrow through the target ring.

• NOTE: The best project gets an automatic 100%.

NOTES:

CATEGORIES: Accuracy, Distance, Rubber Bands, Scavengers, Throwers

MISSION: Throwing Money Away

BRIEF: You and your team have been selected to make a device that throws a coin as far as possible using a clothespin.

MISSION RULES:

1. You will design a device that throws a coin as far as possible, using a clothespin as your spring mechanism.

2. Your primary building materials should be a clothespin and a small cup or holder to hold the coin. You also need a base to hold the entire device.

3. The clothespin must be attached to some sort of base or platform.

4. You will work with one or two partners. Teams may be of no more than 3 people.

5. Up to three tests will be made.

TEACHER'S NOTES: After 3 tests, honors can be given for the best average and/or the highest single shot.

QUIZ GRADE:

A blueprint design of your idea

- Sketch 25%

- Sketch is labeled 25%

- Explanation of strategy 25%

- Conclusions and reflections based on your results 25%

TEST GRADE:

Your completed design and the results of the test.

- Project Completed = 50%

- 50% of your grade depends on how far your project sends the coin.

- NOTE: The best project gets an automatic 100%.

NOTES:

CATEGORIES: Clothespins, Coins, Distance, Throwers

MISSION: Trebuchets

BRIEF: You and your team have been selected to make a device that throws a marble as far as possible using a lever and a counterweight.

MISSION RULES:

1. You will design a device that throws a marble as far as possible, using a lever and a counterweight as your spring mechanism.

2. Your device may be of any dimensions under 12 inches in any one direction.

3. You may build your device from any approved materials found at school or at home.

4. The device must be freestanding and not attached to any surface.

5. You will work with one or two partners. Teams may be of no more than 3 people.

6. Up to three tests will be made. Distance will be measured at the first bounce.

TEACHER'S NOTES: After 3 tests, honors can be given for the best average and/or the highest single shot.

Shooting into a long sand pit allows the ball to stop, rather than bounce. Golf balls could be used, too.

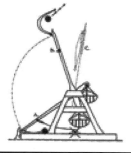

QUIZ GRADE:

A blueprint design of your idea

- Sketch 25%

- Sketch is labeled 25%

- Explanation of strategy 25%

- Conclusions and reflections based on your results 25%

TEST GRADE:

Your completed design and the results of the test.

- Project Completed = 50%

- 50% of your grade depends on how far your project sends the marble.

- NOTE: The best project gets an automatic 100%.

NOTES:

CATEGORIES: Distance, Levers, Marbles, Scavengers, Throwers

150 STEM Labs

SCIENCE * TECHNOLOGY * ENGINEERING * MATH

Towers Projects

MISSION: Claymore Tower

BRIEF: You and your team have been selected to make a the tallest tower possible from clay.

MISSION RULES:

1. You will design a tower.

2. Your device may be any dimensions, but it must be as tall as possible. It must have its own base.

3. You will work with one or two partners. Teams may not be of more than 3 people.

4. You may use any shaping tools you need, but you will only get a specific amount of clay.

5. The tower may not be braced against or attached to any other objects, except for whatever table surface it is built on. Otherwise, it must be completely freestanding.

QUIZ GRADE:

A blueprint design of your idea

- Sketch 25%

- Sketch is labeled 25%

- Explanation of strategy 25%

- Conclusions and reflections based on your results 25%

TEST GRADE:

Your completed design and the results of the test.

- Project Completed = 50%

- 50% of your grade depends on how tall your project is compared to the other group's projects. The projects that do best will get more points.

NOTES:

CATEGORIES: Clay, Height, Towers

MISSION: Foilty Towers

BRIEF: You and your team have been selected to make a the tallest tower possible from a 3 foot length of foil and NOTHING else.

MISSION RULES:

1. You will design a tower.

2. Your device may be any dimensions, but it must be as tall as possible.

3. You will work with one partner. Teams may not be of more than 2 people.

4. You may use only a 3 foot length of aluminum foil and nothing else!

5. The tower may not be braced against or attached to any other objects. It must be completely freestanding.

QUIZ GRADE:

A blueprint design of your idea

- Sketch 25%

- Sketch is labeled 25%

- Explanation of strategy 25%

- Conclusions and reflections based on your results 25%

TEST GRADE:

Your completed design and the results of the test.

- Project Completed = 50%

- 50% of your grade depends on how tall your project is compared to the others.

- NOTE: The tallest project scores an automatic 100%

NOTES:

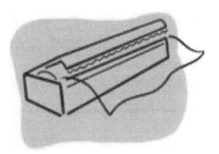

CATEGORIES: Foil, Height, Towers

MISSION: Gummy Towers

BRIEF: You and your team have been selected to make as tall of a tower as possible using only gumdrops and toothpicks.

MISSION RULES:

1. You will research towers and get ideas for a concept for your tower design.

2. Your tower must be as tall as possible using the provided materials.

3. You will work with one or two partners. Teams may not be of more than 3 people.

4. You must only use gumdrops and toothpicks for your project. You teacher will determine how many materials you may use.

5. The tower must be freestanding and may not be attached to any surface.

6. The tower must not collapse for at least 10 seconds when it is being tested.

QUIZ GRADE:

A research paper on towers.

- 2-4 pictures of towers 25%

- A concept idea based on your tower pictures 50%

- Conclusions and reflections based on your results 25%

TEST GRADE:

Your completed design and the results of the test.

- Project Completed = 50%

- 50% of your grade depends on how tall your project is compared to the other group's projects. The projects that do best will get more points.

NOTES:

CATEGORIES: Gumdrops, Height, Toothpicks, Towers

MISSION: House of Cards

BRIEF: You and your team have been selected to make a the tallest tower possible from 6 index note cards, and NOTHING else.

MISSION RULES:

1. You will design a tower.

2. Your device may be any dimensions, but it must be as tall as possible.

3. You will work with one partner. Teams may not be of more than 2 people.

4. You may use only 6 note cards and nothing else. That means no glue, tape, or other adhesives!

5. The tower may not be braced against or attached to any other objects. It must be completely freestanding.

QUIZ GRADE:

Create a blueprint designs for your ideas

- Sketch 25%

- Sketch is labeled 25%

- Explanation of strategies 25%

- Conclusions and reflections based on your results 25%

TEST GRADE:

Your completed design and the results of the test.

- Project Completed = 50%

- 50% of your grade depends on how tall your project is compared to the other group's projects. The projects that are tallest will get more points.

NOTES:

CATEGORIES: Height, Notecards, Towers

MISSION: High Clips

BRIEF: You and your team have been selected to make the tallest tower possible from paper clips!

MISSION RULES:

1. You will design and build a paper clip tower. Your teacher will decide on your maximum number of paperclips.

2. Your tower must be built from only paper clips, although you may use tools like needle-nose pliers or other approved tools to help shape and form your clips.

3. The tower must be freestanding. It will sit on a flat surface and cannot come in contact with walls or other things that might brace it up.

4. You must work with only 1 partner, or alone. Teams ma y not be made up of more than 2 people.

QUIZ GRADE:

Create a blueprint designs for your ideas

- Sketch 25%

- Sketch is labeled 25%

- Explanation of strategies 25%

- Conclusions and reflections based on your results 25%

TEST GRADE:

Your completed design and the results of the test.

- Project Completed = 50%

- 50% of your grade depends on how much tall your project is compared to others.

- NOTE: The tallest project gets an automatic 100%.

NOTES:

CATEGORIES: Height, Paper Clips, Towers

MISSION: King of Cups

BRIEF: You and your team have been selected to make a the tallest tower possible from only plastic cups!

MISSION RULES:

1. You will design a tower.

2. Your device may be any dimensions, but it must be as tall as possible.

3. You will work with one or two partners. Teams may not be of more than 3 people.

4. You may use only plastic cups to build your project. Your teacher will determine your maximum amount of materials.

5. The tower may not be braced against or attached to any other objects, except for whatever table surface it is built on. Otherwise, it must be completely freestanding.

6. Your tower must stand on its own for at least 10 seconds while it is being measured.

TEACHER'S NOTES: A set number, like 10, 12, or 15 cups is suggested.

QUIZ GRADE:

A blueprint design of your idea

- Sketch 25%

- Sketch is labeled 25%

- Explanation of strategy 25%

- Conclusions and reflections based on your results 25%

TEST GRADE:

Your completed design and the results of the test.

- Project Completed = 50%

- 50% of your grade depends on how tall your project is compared to the other group's projects. The projects that do best will get more points.

NOTES:

CATEGORIES: Cups, Height, Towers

MISSION: Leaning Tower of Pasta

BRIEF: You and your team have been selected to make a the tallest tower possible from spaghetti noodles and marshmallows.

MISSION RULES:

1. You will design a tower.

2. Your device may be any dimensions, but it must be as tall as possible. It must have its own base.

3. You will work with one or two partners. Teams may not be of more than 3 people.

4. You may use only spaghetti and marshmallows to build your project. Your teacher will determine your maximum amount of materials,

5. The tower may not be braced against or attached to any other objects, except for whatever table surface it is built on. Otherwise, it must be completely freestanding.

QUIZ GRADE:

A blueprint design of your idea

- Sketch 25%

- Sketch is labeled 25%

- Explanation of strategy 25%

- Conclusions and reflections based on your results 25%

TEST GRADE:

Your completed design and the results of the test.

- Project Completed = 50%

- 50% of your grade depends on how tall your project is compared to the other group's projects. The projects that do best will get more points.

NOTES:

CATEGORIES: Height, Marshmallows, Pasta, Towers

MISSION: Leaning Towers

BRIEF: You and your team have been selected to make a the tallest tower possible from 3 pieces of paper and a foot of tape.

MISSION RULES:

1. You will design a tower.

2. Your device may be any dimensions, but it must be as tall as possible. It must have its own base.

3. You will work with two or three partners. Teams may not be of more than 4 people.

4. You may use only scissors, exactly one foot of tape or less, and 3 sheets of copy paper.

5. The tower may not be braced against or attached to any other objects. It must be completely freestanding.

QUIZ GRADE:

A blueprint design of your idea

- Sketch 25%

- Sketch is labeled 25%

- Explanation of strategy 25%

- Conclusions and reflections based on your results 25%

TEST GRADE:

Your completed design and the results of the test.

- Project Completed = 50%

- 50% of your grade depends on how tall your project is compared to the other group's projects. The projects that do best will get more points.

- NOTE: There is a -5% penalty for each new sheet of paper you are exchanging for one your team made a mistake on and wasted.

NOTES:

CATEGORIES: Height, Paper, Towers

MISSION: Peanut Tower

BRIEF: You and your team have been selected to make as tall of a tower as possible using only foam packing peanuts and toothpicks.

MISSION RULES:

1. You will research towers and get ideas for a concept for your tower design.

2. Your tower must be as tall as possible using the provided materials.

3. You will work with one or two partners. Teams may not be of more than 3 people.

4. You must only use packing peanuts and toothpicks for your project. You teacher will determine how many materials you may use.

5. The tower must be freestanding and may not be attached to any surface.

6. The tower must not collapse for at least 10 seconds when it is being tested.

QUIZ GRADE:

A research paper on towers.

- 2-4 pictures of towers 25%

- A concept idea based on your tower pictures 50%

- Conclusions and reflections based on your results 25%

TEST GRADE:

Your completed design and the results of the test.

- Project Completed = 50%

- 50% of your grade depends on how tall your project is compared to the other group's projects. The projects that do best will get more points.

NOTES:

CATEGORIES: Height, Packing Peanuts, Toothpicks, Towers

MISSION: The Pipeworks

BRIEF: You and your team have been selected to make a the tallest tower possible from pipe cleaners.

MISSION RULES:

1. You will design a tower.

2. Your device may be any dimensions, but it must be as tall as possible. It must have its own base.

3. You will work with one or two partners. Teams may not be of more than 3 people.

4. You may use only scissors and a specific amount of pipe cleaners to build your tower.

5. The tower may not be braced against or attached to any other objects. It must be completely freestanding.

QUIZ GRADE:

A blueprint design of your idea

- Sketch 25%

- Sketch is labeled 25%

- Explanation of strategy 25%

- Conclusions and reflections based on your results 25%

TEST GRADE:

Your completed design and the results of the test.

- Project Completed = 50%

- 50% of your grade depends on how tall your project is compared to the other group's projects. The projects that do best will get more points.

- NOTE: There is a -5% penalty for each new sheet of paper you are exchanging for one your team made a mistake on and wasted.

NOTES:

CATEGORIES: Height, Pipe Cleaners, Towers

MISSION: Starchy Goodness

BRIEF: You and your team have been selected to make a the tallest tower possible from only glue and string!

MISSION RULES:

1. You will design a tower.

2. Your device may be any dimensions, but it must be as tall as possible. It must have its own base.

3. You will work with one or two partners. Teams may not be of more than 3 people.

4. You may use only string/yarn and glue to build your project. Your teacher will determine your maximum amount of materials.

5. The tower may not be braced against or attached to any other objects, except for whatever table surface it is built on. Otherwise, it must be completely freestanding.

TEACHER'S NOTES: It is suggested that you use wax paper to set up 'cables' of thread or yarn. Glue will help them set into a specific shape when it dries. Then they can be assembled into a tower.

QUIZ GRADE:

A blueprint design of your idea

- Sketch 25%

- Sketch is labeled 25%

- Explanation of strategy 25%

- Conclusions and reflections based on your results 25%

TEST GRADE:

Your completed design and the results of the test.

- Project Completed = 50%

- 50% of your grade depends on how tall your project is compared to the other group's projects. The projects that do best will get more points.

NOTES:

CATEGORIES: Glue, Height, String, Towers, Yarn

MISSION: Toothpick Towers

BRIEF: You and your team have been selected to make a 36 inch tower made from only glue and toothpicks.

MISSION RULES:

1. You will design a tower.

2. Your device may be any dimensions, but it must be at least 36 inches tall.

3. You will work with two or three partners. Teams may not be of more than 4 people.

4. You may use only glue and toothpicks in the final project, as well as wax paper as a working surface to help things not stick as they dry. (there may be some bracing used when the project's glue is drying, but they must not be present in the finished project).

5. The tower may not be braced against or attached to any other objects. It must be completely freestanding.

6. No more than 6 inches at the top of the tower may be made from a needle-like spike.

QUIZ GRADE:

A blueprint design of your idea

• Sketch 25%

• Sketch is labeled 25%

• Explanation of strategy 25%

• Conclusions and reflections based on your results 25%

TEST GRADE:

Your completed design and the results of the test.

• Project Completed = 50%

• Tower is 36" or above = +50%

• Tower is 30" or above = +40%

• Tower is 24" or above = +30%

• Tower is 18" or above = +20% and anything else is failing.

NOTES:

CATEGORIES: Height, Toothpicks, Towers

MISSION: Tube Frame Towers

BRIEF: You and your team have been selected to make as tall of a tower as possible using only a dozen plastic straws.

MISSION RULES:

1. You will design a tower.

2. Your device may be any dimensions, but it must only use straws.

3. You will work with one or two partners. Teams may not be of more than 3 people.

4. The tower may not be braced against or attached to any other objects. It must be completely freestanding.

QUIZ GRADE:

A blueprint design of your idea

- Sketch 25%

- Sketch is labeled 25%

- Explanation of strategy 25%

- Conclusions and reflections based on your results 25%

TEST GRADE:

Your completed design and the results of the test.

- Project Completed = 50%

- 50% of your grade depends on how tall your project is compared to the others.

- NOTE: The tallest project scores an automatic 100%

NOTES:

CATEGORIES: Height, Plastic Straws, Towers

150 STEM Labs

SCIENCE * TECHNOLOGY * ENGINEERING * MATH

Tracks Projects

MISSION: Copper Road

BRIEF: You and your team have been selected to make the longest, trickiest course in which to deliver a penny to its final destination. The longer and trickier the better.

MISSION RULES:

1. You will design a roller coaster for a penny out of tape, paper, card stock, other supplied materials, and materials you can find at home that are approved for use.

2. The roller coaster must successfully deliver 3 pennies, one at a time, to a collection point at the base of the project. Each penny that fails to make it to the end of the course will result in a penalty for the total score.

3. You may test at home. In fact, you're encouraged to test at home! Assembly and design may also take place at school, but time is limited.

4. Teams may be of no more than 4 people.

5. Suggested Tricks to include are: stairs, vertical loops, jumps, horizontal spirals, switchbacks, tunnels, funnels, trap doors, and drain pans. Be original designing tricks. Tricks can be anything where the penny just isn't rolling in a straight path.

TEACHER'S NOTES: Some hints should be given about the way a penny rolls and slides that can be translated into designs.

QUIZ GRADE:

A decorated advertisement poster that explains your concept, names your roller coaster, lists the features of the coaster, and explains which team member was responsible for each part.

• Coaster Name 10%

• Coaster Concept 25%

• List of Features 25%

• Sketch/Artwork 30%

• Who did What? 10%

TEST GRADE:

Your completed design and the results of the test.

• Project Completed = 25%

• 1% per second of average time (average of all 3 tests)

• -5% for each penny that fails to make it to the end of the track. You will get 1 restart or 'nudge' per penny for free, but then it costs you.

• +10% per different trick, or +5% for a second trick of the same kind already used. No points for more one of the same kind of trick.

CATEGORIES: Coins, Paper, Time, Tracks

MISSION: Down the Chute

BRIEF:

You and your team have been selected to design a delivery device that can take a marble 5 foot across the room and drop it into an open and empty soda bottle.

MISSION RULES:

1. You will design a delivery system to transport a marble 5 feet across the room into a bottle.

2. Your finished project must be built of only paper, card stock, tape, glue, and other materials approved by your teacher.

3. You will work with one or two partners. Teams may be of no more than 3 people.

4. Your project may NOT be attached to the bottle in any way.

5. Your project must be freestanding, not attached to the floor or any furniture.

QUIZ GRADE:

Create a blueprint design for your ideas

- Sketch 25%

- Sketch is labeled 25%

- Explanation of strategies 25%

- Conclusions and reflections based on your results 25%

TEST GRADE:

Your completed design and the results of the test.

- Project Completed = 40%

50% of your grade depends on your success in your 3 trials.

- Success 3/3 = +50%,

- Success 2/3 = +35%

- Success 1/3 = +20%, otherwise no points.

NOTES:

CATEGORIES: Accuracy, Marbles, Tracks

MISSION: The Irrigator

BRIEF: You and your team have been selected to make the longest, trickiest course in which to deliver water to its final destination. The longer and trickier the better.

MISSION RULES:

1. You will design a roller coaster for water made out of tape, plastic, clay, tubing, straws, styrofoam, wood, and materials you can find at home that are approved for use.

2. The roller coaster must successfully deliver 1 cup of water to a collection point at the base of the project. Water that leaks or fails to make it to the collection point will cost you points.

3. You may test at home. In fact, you're encouraged to test at home! Assembly and design may also take place at school, but time is limited.

4. Teams may be of no more than 4 people.

5. Suggested Tricks to include are: stairs, vertical loops, jumps, horizontal spirals, switchbacks, tunnels, funnels, trap doors, and drain pans. Be original designing tricks. Tricks can be anything where the water just isn't rolling in a straight path.

TEACHER'S NOTES: Adding color to the water makes it easier to track leaks and adds excitement. Maybe use red for trial 1, blue for trial 2, and green for trial 3.

QUIZ GRADE:

A decorated advertisement poster that explains your concept, names your roller coaster, lists the features of the coaster, and explains which team member was responsible for each part.

• Coaster Name 10%

• Coaster Concept 25%

• List of Features 25%

• Sketch/Artwork 30%

• Who did What? 10%

TEST GRADE:

Your completed design and the results of the test.

• Project Completed = 25%

• 5% per second of average time (average of all 3 tests)

• -5% for each ounce of water that fails to make it through the track.

• +10% per different trick, or +5% for a second trick of the same kind already used. No points for more one of the same kind of trick.

CATEGORIES: Plastic Straws, Scavengers, Time, Tracks, Water

MISSION: Marble Madness

BRIEF: You and your team have been selected to make the longest, trickiest course in which to deliver a marble to its final destination. The longer and trickier the better.

MISSION RULES:

1. You will design a roller coaster for a marble out of tape, paper, card stock, other supplied materials, and materials you can find at home that are approved for use.

2. The roller coaster must successfully deliver 3 marbles, one at a time, to a collection point at the base of the project. Each marble that fails to make it to the end of the course will result in a penalty for the total score.

3. You may test at home. In fact, you're encouraged to test at home! Assembly and design may also take place at school, but time is limited.

4. Teams may be of no more than 4 people.

5. Suggested Tricks to include are: stairs, vertical loops, jumps, horizontal spirals, switchbacks, tunnels, funnels, trap doors, and drain pans. Be original designing tricks. Tricks can be anything where the ball just isn't rolling in a straight path.

QUIZ GRADE:

A decorated advertisement poster that explains your concept, names your roller coaster, lists the features of the coaster, and explains which team member was responsible for each part.

- Coaster Name 10%

- Coaster Concept 25%

- List of Features 25%

- Sketch/Artwork 30%

- Who did What? 10%

TEST GRADE:

Your completed design and the results of the test.

- Project Completed = 25%

- 1% per second of average time (average of all 3 marble tests)

- -5% for each marble that fails to make it to the end of the track. You will get 1 restart or 'nudge' per marble for free, but then it costs you.

- +10% per different trick, or +5% for a second trick of the same kind already used. No points for more one of the same kind of trick.

CATEGORIES: Marbles, Scavengers, Time, Tracks

MISSION: Ping Pong Madness

BRIEF: You and your team have been selected to make the longest, trickiest course in which to deliver a ping pong ball to its final destination. The longer and trickier the better.

MISSION RULES:

1. You will design a roller coaster for a ping pong ball out of tape, cardboard tubes, paper, card stock, other supplied materials, and materials you can find at home that are approved for use.

2. The roller coaster must successfully deliver 3 ping pong balls, one at a time, to a collection point at the base of the project. Each ball that fails to make it to the end of the course will result in a penalty for the total score.

3. You may test at home. In fact, you're encouraged to test at home! Assembly and design may also take place at school, but time is limited.

4. Teams may be of no more than 4 people.

5. Suggested Tricks to include are: stairs, vertical loops, jumps, horizontal spirals, switchbacks, tunnels, funnels, trap doors, and drain pans. Be original designing tricks. Tricks can be anything where the ball just isn't rolling in a straight path.

QUIZ GRADE:

A decorated advertisement poster that explains your concept, names your roller coaster, lists the features of the coaster, and explains which team member was responsible for each part.

- Coaster Name 10%

- Coaster Concept 25%

- List of Features 25%

- Sketch/Artwork 30%

- Who did What? 10%

TEST GRADE:

Your completed design and the results of the test.

- Project Completed = 25%

- 1% per second of average time (average of all 3 ping pong balls tests)

- -5% for each ping pong ball that fails to make it to the end of the track. You will get 1 restart or 'nudge' per ball for free, but then it costs you.

- +10% per different trick, or +5% for a second trick of the same kind already used. No points for more one of the same kind of trick.

CATEGORIES: Cardboard Tubes, Paper, Ping Pong Balls, Time, Tracks

MISSION: Ski Jump

BRIEF: You and your team have been selected to make a ramp to jump marbles, golf balls, and/or ping pong balls as far as possible.

MISSION RULES:

1. You will design a jump to help a ball (marbles/ping pong balls/golf balls) as far as possible.

2. Distance is only measured to the first bounce. Rolling distance is not counted.

3. Your finished project must be built of cardboard tubes, tape, glue, and other materials approved by your teacher.

4. You will work with one or two partners. Teams may be of no more than 3 people.

5. Your project must be freestanding, and may not be attached to any surface.

TEACHER'S NOTES: You might want to have students start collecting cardboard tubes from paper towels and gift wrapping paper a good deal ahead of time before they start the project. Toilet paper rolls are likely not sanitary...

QUIZ GRADE:

Create a blueprint design for your ideas

- Sketch 25%
- Sketch is labeled 25%
- Explanation of strategies 25%
- Conclusions and reflections based on your results 25%

TEST GRADE:

Your completed design and the results of the test.

- Project Completed = 50%
- 50% of your grade depends on how far the ball flies before hitting the ground compared to the other group's projects. The projects that do best will get more points.

NOTE: If you are testing more than 1 type of ball in your design, high averages or best distance per type may receive extra points.

NOTES:

CATEGORIES: Cardboard Tubes, Distance, Golf Balls, Marbles, Ping Pong Balls

MISSION: Take the High Road

BRIEF: You and your team have been selected to design a racetrack that takes at least 10 seconds for a matchbox or hot wheels car to complete.

MISSION RULES:

1. You will design a track from scavenged supplies, including: tape, straws, paper clips, note cards, cardboard tubes/boxes, paper, card stock, and other approved supplies.

2. Your finished tracks must fit within the classroom, or within a workspace designated for your team by your teachers.

3. You will work with two or three partners. Teams may be of no more than 4 people.

4. Time will be measured with a stopwatch as your car races down the track. You will be given 3 attempts to get the longest time. possible.

QUIZ GRADE:

A blueprint design of your idea

- Sketch 25%

- Sketch is labeled 25%

- Explanation of strategy 25%

- Conclusions and reflections based on your results 25%

TEST GRADE:

Your completed design and the results of the test.

- Project Completed = 50%

- 50% of your grade depends on how long it takes for your project to complete the track.

- NOTE: There is a -5% penalty for every push or restart required, but there may be bonus % added for tricks like jumps, loops, or otherwise.

NOTES:

CATEGORIES: Cars, Scavengers, Time, Tracks

MISSION: Tubular Balls

BRIEF: You and your team have been selected to design a delivery device that can take a ping pong ball across the room and drop it into an open container.

MISSION RULES:

1. You will design a delivery system to transport a ping pong ball as far as possible across the room into an open container.

2. Your finished project must be built of only paper, card stock, cardboard tubes, tape, glue, and other materials approved by your teacher.

3. You will work with one or two partners. Teams may be of no more than 3 people.

4. Your project may NOT be attached to the open container in any way.

5. Your project may use furniture as a pivot or fulcrum, but it must not be taped to the floor or furniture.

6. Your device may not be an enclosed tube for more than 50% of its length.

QUIZ GRADE:

Create a blueprint design for your ideas

- Sketch 25%

- Sketch is labeled 25%

- Explanation of strategies 25%

- Conclusions and reflections based on your results 25%

TEST GRADE:

Your completed design and the results of the test.

- Project Completed = 25%

- 50% of your grade depends on your success in your 3 trials. Success 3/3 = +50%, Success 2/3 = +35%, and Success 1/3 = +20%, otherwise no points.

- The final 25% of your grade is the comparative length of your project compared to the others. Longer projects that are successful get more points.

- NOTE: The longest project that deposits all 3 balls into the container gets an automatic 100%

NOTES:

CATEGORIES: Accuracy, Cardboard Tubes, Distance, Ping Pong Balls, Tracks

MISSION: Wooden Railway

BRIEF: You and your team have been selected to make a train out of wood and glue!

MISSION RULES:

1. You will design a rolling train of at least 3 cars made entirely from wood and glue.

2. The train must run on a popsicle stick railway at least 18 inches long with at least one curve in it.

3. You will work with two or three partners. Teams may be of no more than 4 people.

4. Suggested wooden materials are: wooden thread spools (plastic may be substituted if wooden ones are not available), toothpicks, popsicle sticks, and dowel rods.

TEACHER'S NOTES: Wax paper is suggested as a non-sticky surface for glued pieces to dry upon.

QUIZ GRADE:

A blueprint design of your idea

- Sketch 25%

- Sketch is labeled 25%

- Explanation of strategy 25%

- Conclusions and reflections based on your results 25%

TEST GRADE:

Your completed design and the results of the test.

- Project Completed = 50%

- 50% of your grade depends on how well your train moves.

- NOTE: There is a 10% penalty for each restart or time the car(s) come off the track.

NOTES:

CATEGORIES: Glue, Popsicle Sticks, Spools, Toothpicks, Tracks, Trains, Wood

ANDREW FRINKLE

BRIEF: A quick look at the author, illustrator, game designer, and creator of over 140 titles!

ABOUT THE AUTHOR:

Andrew Frinkle is an award-nominated teacher and writer with experience in America and overseas, as well as years developing educational materials for big name educational sites like Have Fun Teaching. He has taught PreK all the way up to adult classes, and has focused on ESOL/EFL techniques, as well as STEM Education. With two young children at home now, he's been developing more and more teaching strategies and books aimed at helping young learners.

Andrew Frinkle is the founder & owner of MediaStream Press LLC. Learn more at www.MediaStreamPress.com, which also maintains the following educational websites:
www.50STEMLabs.com
www.common-core-assessments.com
www.littlelearninglabs.com

Additionally, Andrew also writes fantasy and science fiction novels under the pen name Velerion Damarke and writes/illustrates children's fiction as Andrew Frinkle. Find out more at:
www.AndrewFrinkle.com
www.underspace.org

SNAZZY PHOTO:

EMAIL ME:

mediastreampress@gmail.com

CATEGORIES: Author, Educator, Game Designer, Illustrator, Musician

150 STEM LABS

I hope you enjoyed this volume! Make sure to check out other volumes in the series and other books at www.50STEMLabs.com

MEDIASTREAM PRESS

WWW.50STEMLABS.COM

Made in the USA
Lexington, KY
11 August 2018